Spon's Estimating Costs Guide to Finishings

Painting, decorating, plastering and tiling

Second Edition

Bryan Spain

CRC Press
Taylor & Francis Group
Boca Raton   London   New York

CRC Press is an imprint of the
Taylor & Francis Group, an **informa** business

CRC Press
Taylor & Francis Group
6000 Broken Sound Parkway NW, Suite 300
Boca Raton, FL 33487-2742

© 2005 by Taylor & Francis Group, LLC
CRC Press is an imprint of Taylor & Francis Group, an Informa business

No claim to original U.S. Government works

ISBN 13: 978-0-415-43443-0 (pbk)

Visit the Taylor & Francis Web site at
http://www.taylorandfrancis.com

and the CRC Press Web site at
http://www.crcpress.com

Contents

External work

Preface

This is the second edition of Spon's Estimating Costs Guide to Finishings covers plastering, tiling, painting work. It is intended to provide accurate cost data to enable small plastering and painting firms to prepare estimates and quotations more quickly and more accurately.

This need for speed and accuracy is vital for all contractors operating in the competitive domestic construction market. Most contractors have the skills necessary to carry out the work together with the capacity for dealing with the setbacks that are part of the normal construction process. But they rarely have enough time to complete the many tasks that must be carried out in order to trade profitably. This book aims to help contractors to prepare their estimates by providing thousands of rates for small work involving finishings. If used sensibly, the book can save valuable time for contractors in the preparation of their bids.

I have received a great deal of support in the research necessary for this type of book and I am grateful to those individuals and firms who have provided the cost data and other information. In particular, I am indebted to Mark Loughrey of Loughrey & Co. Ltd, Chartered Accountants of Hoylake (tel: 0151-632 3298 or mjloughrey@accountant.com), who are specialists in advising small construction businesses. Their research for the information in the business section is based on tax legislation in force in April 2007.

Although every care has been taken in the preparation of the book, neither the Publisher nor I can accept any responsibility for the use of the information provided by any firm or individual. Finally, I would welcome any constructive criticism of the book's contents and suggestions that could be incorporated into future editions.

Bryan Spain
spainandpartners@btconnect.com
April 2007

Standard Method of Measurement/trades link

The contents of this book are presented under trade headings and the following table provides a link to the Standard Method of Measurement (SMM7).

Plastering and tiling

M10 Sand cement/granolithic screeds
M20 Plastered coatings
M30 Metal mesh lathing
K10 Plasterboard dry lining
M40 Quarry/ceramic tiling
M50 plastic/lino tiling

Painting and wallpapering

M60 Painting/clear finishing
M52 Decorative papers
Wallpapering

Introduction

This edition of Spon's Estimating Costs Guide to Finishings follows the layout, style and contents of other books in this series. The contents of the book cover unit rates, project costs, repairs, tool and equipment hire, general advice on business matters and other information useful to those involved in the commissioning and construction of landscaping and external works. The unit rates section presents analytical rates for work up to about £50,000 in value and the business section covers advice on starting and running a business together with information on taxation and VAT matters.

Materials

In the domestic construction market, contractors are not usually able to purchase materials in large quantities and cannot benefit from the discounts available to larger contractors. An average of 10% to 15% discount has been allowed on normal trade prices.

Labour

The hourly labour rates for craftsmen and general operatives are based upon the current wage awards. These are set at:

Craftsman	£15.00
General operative	£12.00

These rates include provision for NIC Employers' contribution, CITB levy, insurances, public and annual holidays, severance pay and tool allowances where appropriate.

Headings

The following column headings have been used.

Unit	Labour	Hours £	Materials £	O & P £	Total £
m2	0.20	3.00	2.20	0.78	5.98

Unit

This column shows the unit of measurement for the item description:

nr	number
m	linear metre
m2	square metre
m3	cubic metre.

Labour

In the example shown, 0.20 represents the estimated time estimated to carry out one square metre of the described item, i.e. 0.20 hours.

Hours

The entry of £3.00 is calculated by multiplying the entry in the Labour column by the labour rate of £15.00.

Materials

This column displays the cost of the materials required to carry out one square metre of the described item, i.e. £2.20.

O & P (Overheads and profit)

This has been set at 15% and is deemed to cover head office and site overheads including:

- heating
- lighting
- rent
- rates
- telephones
- secretarial services
- insurances
- finance charges
- transport
- small tools
- ladders
- scaffolding etc.

Total

This is the total of the Hours, Materials and Overheads and Profit columns.

Contracting

Tradesmen and small contractors can act as main contractors (working for a client direct) or as a subcontractor working for another contractor. Although a contract exists between a subcontractor and a main contractor, there is no contractual link between a subcontractor and an Employer.

In general terms this means that the subcontractor cannot make any claims against the Employer direct and vice-versa. It also means that the subcontractor should not accept any instructions from the Employer or his representative because this could be taken as establishing a privity of contract between the two parties,

A subcontractor must be aware of his role in the programme because if he causes a contractor to overrun the completion date for the main contract he may become liable for the full amount of liquidated damages on the main contract plus the cost of damages that the contractor and other subcontractors may have suffered.

A well-organised subcontractor will keep a full set of daily site records, staffing levels, plant on site, weather charts and such like. It also cannot be over emphasised that any verbal instructions that the subcontractor receives, should be confirmed immediately to the contractor in writing with the name of the person who issued them.

This procedure is extremely important because it may eventually save the subcontractor considerable expense if someone tries to lay the blame for delays to the contract at his door. It is also important that instructions should only be taken from the contractor and he should be informed if another party attempts to do so.

Contractor's discount

Most sub-contracts allow for a discount to the contractor of 2½ % from the subcontractor's account. This means that the subcontractor must add this discount to his prices by adding 1/39th to his net rates.

Payment and retention

Payment is normally made on a monthly basis. The subcontractor should submit his account to the contractor who then incorporates it into his own payment application and passes it on to the Architect or Employer's representative for certification of payment. When the subcontractor receives his payment it will be reduced by 5% retention.

This money is held by the Employer and will be released in two parts. The first part, or moiety, is paid at the completion of work and, in the subcontractor's case, this may be either when he has finished his work or when the contractor has completed the contract as a whole (known as practical completion) depending upon the contract conditions. The second part is released at the end of the defects liability period.

This money is held by the Employer and will be released in two parts. The first part, or moiety, is paid at the completion of work and, in the subcontractor's case, this may be either when he has finished his work or when the contractor has completed the contract as a whole (known as practical completion) depending upon the contract conditions. The second part is released at the end of the defects liability period.

Defects liability period

This is the period of time (normally 12 months) during which the sub-contractor is contractually bound to return to the job to rectify any mistakes or bad pieces of workmanship. This could either be twelve months from when he completes his work or twelve months from when the main contract is completed depending upon the wording of the sub contract.

Period for completion

Usually, a sub-contractor will be given a period of time in which he must complete the work and he must ensure that he has the capability to do the work within that period. Failure to meet the agreed completion date could have serious consequences.

Under certain circumstances, however, particularly with nominated sub-contracts, the sub-contractor may be requested to state the period of time he requires to do the work. If this is the case, then careful thought must be given to the time inserted. Too short a time may put him at financial risk but too long a time may prejudice the opportunity of winning the contract.

Damages for non-completion

A clause is usually inserted within each sub-contract stating that the sub-contractor is liable for the financial losses that contractor suffers due to the sub-contractor's non-completion of work on time. This will include the amount of liquidated and ascertained damages contained within the main contract, together with the contractor's own direct losses and the direct losses of his other sub-contractors. As can be seen, the potential cost to the sub-contractor can be large so he must take care to expedite the work with due diligence to avoid incurring these costs.

Variations

All sub-contracts contain a clause allowing the sub-contract work to be varied without invalidating it. The sub-contractor will normally be paid any additional cost he incurs in carrying out variations.

Insurances

The subcontractor is responsible for insuring against injury to persons or property and against loss of plant and materials. These insurances could be taken out for each individual job, although it is more common to take out blanket policies based on the turnover the firm has achieved in the previous year.

Extensions of time

The subcontractor will normally be entitled to a longer period of time to complete the work if he is delayed or interrupted by reasons beyond his control (known as an extension of time). Most sub-contracts list the reasons and in some cases the sub-contractor may also be entitled to additional monies as well as an extension of time.

Domestic sub-contracts

In domestic sub-contracts the contractor would obtain competitive quotations from various subcontractors of his own choice and these may be based on a bill of quantities, specification and drawings, or schedules of work. Accompanying the enquiry should also be a form of sub-contract that the subcontractor will be required to complete.

There are several points that may affect costs and which the subcontractor should bear in mind. These are

1. Whether the rates and prices are to include for any contractor's discount (normally expressed as plus 1/39th to allow 2½%).

2. Whether the contractor is to supply any labour or plant to assist the subcontractor in either carrying out any of the work or in off-loading materials.

3. What facilities (if any) the contractor will provide for the subcontractor such as mess rooms, welfare facilities, office accommodation and storage facilities.

4. Whether the contractor is to dispose of the subcontractor's rubbish.

Contracting

Often a subcontractor will find himself working under a private contract, written or implied. This usually takes the form of working for a domestic householder or a small factory owner and the following procedures usually apply in this type of work.

Estimate

The initial approach would usually come from a purchaser, e.g. 'How much will it cost to have my garden landscaped?' At this stage, he may only want an approximate cost in order to see if he can afford to have the work carried out as opposed to a quotation which is a firm offer to do the work. Therefore, a brief description of the work to be carried out together with an approximate price will suffice.

However, it should be made clear that the price is an estimate and does not constitute an offer that may be accepted by the purchaser. The estimate may be based on a telephone conversation only, e.g. 'It will cost about £4,000 to £ 5,000 to landscape your garden', or it could be based on a brief visit to the house. In either case, little time should be spent on an estimate and it is generally wise to express it as a price range.

Quotation

A quotation is generally seen as an offer to do the work for the price quoted, and could constitute a simple contract if accepted. It follows that some time and effort should be spent in compiling a quotation to save arguments at a later stage. One should always remember that the contractor is the expert and must use his expertise in order to guide the purchaser and should discuss the work with him in full. He should tell the purchaser exactly what he is getting for the price and also what he is not.

This may mean going in to some detail such as what will happen to the surplus excavated materials, how access will be gained, how long the job will take and similar items.

The contractor should also find out from the purchaser exactly what restrictions (if any) will be placed upon him. For instance, will the purchaser keep the drive clear of cars to allow a skip to be used and will the contractor only be allowed entry to the premises on certain days and/or at certain times? These factors, should be ascertained in advance, and the costs of complying with them should be made known to the purchaser who may decide to take steps to change the restrictions.

Once the contractor has considered all the relevant factors then the formal written quotation can be produced. It should state precisely what the purchaser is getting for his money, including when and how long the job will take and contain all the salient points of discussions that have taken place.

After a quotation has been submitted then all that needs to be done is for the purchaser to accept it. Although a verbal acceptance would constitute a binding agreement, it is always more satisfactory if the acceptance is made in writing.

Payments

There is much debate on how and when payments should be made in domestic situations. Ideally from a contractor's point of view to be paid in advance would be the most advantageous, but the chances of the purchaser wishing to do this are remote.

On the other hand, it may cause undue financial hardship to a recently self-employed contractor to have to buy all the materials himself and not get paid until all the work is completed. Whatever payment policy is adopted it must be agreed with the purchaser in advance and form part of the written quotation.

Possible alternatives are

1. Being paid when the work is complete. This is probably the best method from a public relations aspect and contractors who can complete a job in a few days should have no difficulty in adopting this policy.

2. Being paid before the work is done. This is only really feasible where the contractor concerned is of unquestionable reputation or is well known to the purchaser.

3. Being paid for materials as they are bought and delivered with the balance paid when the work is complete. This could be a practical solution for smaller contractors, but the purchaser will probably want proof of the material costs, so careful handling of invoices is necessary.

4. Some form of stage payments that usually take the form of agreed percentages of the quotation price or agreed parts of the quotation price paid after stages of the work have been carried out.

Pricing and variations

It is important that some method of recording, pricing and being paid for variations is agreed at the outset and this is particularly relevant when dealing with private clients. Unforeseen additions, more than any other item, are the main cause of disputes and are often avoidable.

The risk of this type of dispute can be reduced by ensuring that the original quotation is as detailed as possible. The detailed specification of the materials could be contained within the descriptions or done separately. A quotation broken down in this way is detailed enough to enable the purchaser to ascertain that he is not being overcharged for any variations that may occur and yet is not so detailed that the purchaser is going to question the price of every detail.

Also, if the purchaser should wish to change anything himself then there are no arguments on what was included in the original quotation.

If variations occur, it must be established who should pay for them. There are three main types of variations.

1. Those instructed by the purchaser.

2. Those that should have been included in the original quotation.

3. Those that are necessary due to events that could not have been foreseen.

The liabilities for 1 and 2 are relatively straightforward. If the purchaser says he wants a different paving flag to his original choice, then he must bear the additional cost. Conversely, if the contractor forgot to include the cost of the sub-base in his quotation then it is only fair that he bears the cost.

Item 3 is more difficult. If it is the purchaser who is receiving the benefit of the variations and if they were not foreseeable, then it would be logical to assume that it is the purchaser who should bear the cost. One example would be if a painter discovered some rotting timbers he would obviously not be expected to pay for the work.

Other instances may not be as clear cut as this example and it may become necessary to arrive at a cost-sharing arrangement if genuine doubt exists. Variations should preferably be agreed in advance before the work is carried out. They should be recorded and signed by both parties and, wherever possible, priced in detail and agreed.

Part One

UNIT RATES – PLASTERING AND TILING

Linings and plasterboard

Suspended ceilings

Screeds and granolithic

Plastered coatings

Metal lathing and beads

Floor coverings

Wall tiling

Partitions and dry lining

External render

Repair work

	Unit	Labour Hours	Mat'ls £	O & P £	Total £

LININGS AND PLASTERBOARD

Gypsum tapered edge plasterboard, fixed with nails, joints left
open to receive plaster or similar

	Unit	Labour Hours	Mat'ls £	O & P £	Total £
9.5mm thick to ceilings over 300mm wide	m2	0.30	4.50	2.41	1.04	7.95
9.5mm thick to beams over 600mm wide	m2	0.33	4.95	2.41	1.10	8.46
9.5mm thick to beams less than 600mm wide	m	0.18	2.70	1.34	0.61	4.65
12.5mm thick to ceilings over 300mm wide	m2	0.35	5.25	2.97	1.23	9.45
12.5mm thick to beams over 600mm wide	m2	0.40	6.00	2.97	1.35	10.32
12.5mm thick to beams less than 600mm wide	m	0.20	3.00	1.58	0.69	5.27

Gypsum tapered edge plasterboard, fixed with nails, joints
filled with filler and taped for
direct decoration

	Unit	Labour Hours	Mat'ls £	O & P £	Total £
9.5mm thick to ceilings over 300mm wide	m2	0.35	5.25	2.81	1.21	9.27
9.5mm thick to beams over 600mm wide	m2	0.38	5.70	2.81	1.28	9.79
9.5mm thick to beams less than 600mm wide	m	0.22	3.30	1.50	0.72	5.52
12.5mm thick to ceilings over 300mm wide	m2	0.40	6.00	3.15	1.37	10.52
12.5mm thick to beams over 600mm wide	m2	0.44	6.60	3.15	1.46	11.21
12.5mm thick to beams less than 600mm wide	m	0.25	3.75	1.69	0.82	6.26

	Unit	Labour Hours	Mat'ls £	O & P £	Total £
Tapered edge plank, 19mm thick					
9.5mm thick to walls over 300mm wide	m2	0.34	5.10	4.78	1.48	11.36
9.5mm thick to columns over 600mm wide	m2	0.36	5.40	4.78	1.53	11.71
9.5mm thick to beams less than 600mm wide	m	0.20	3.00	1.95	0.74	5.69
9.5mm thick to sides of openings less than 600mm wide	m	0.22	3.30	1.95	0.79	6.04
12.5mm thick to walls over 300mm wide	m2	0.38	5.70	5.32	1.65	12.67
12.5mm thick to columns over 600mm wide	m2	0.42	6.30	5.32	1.74	13.36
12.5mm thick to beams less than 600mm wide	m	0.22	3.30	2.35	0.85	6.50
12.5mm thick to sides of openings less than 600mm wide	m	0.24	3.60	2.35	0.89	6.84
Gypsum plastic faced wallboard, butt jointed, fixed to softwood					
9.5mm thick to walls over 300mm wide	m2	0.28	4.20	4.66	1.33	10.19
9.5mm thick to columns over 600mm wide	m2	0.30	4.50	4.66	1.37	10.53
9.5mm thick to beams less than 600mm wide	m	0.18	2.70	1.86	0.68	5.24
9.5mm thick to sides of openings less than 600mm wide	m	0.20	3.00	1.86	0.73	5.59
12.5mm thick to walls over 300mm wide	m2	0.32	4.80	5.20	1.50	11.50
12.5mm thick to columns over 600mm wide	m2	0.36	5.40	5.20	1.59	12.19

	Unit	Labour	Hours £	Mat'ls £	O & P £	Total £
12.5mm thick to beams less than 600mm wide	m	0.24	3.60	2.17	0.87	6.64
12.5mm thick to sides of openings less than 600mm wide	m	0.22	3.30	2.17	0.82	6.29

Dry lining system comprising plain grade tapered edge wall board, joints filled with filler and taped for direct decoration

	Unit	Labour	Hours £	Mat'ls £	O & P £	Total £
9.5mm thick to walls over 300mm wide	m2	0.38	5.70	4.66	1.55	11.91
9.5mm thick to columns over 600mm wide	m2	0.40	6.00	4.66	1.60	12.26
9.5mm thick to beams less than 600mm wide	m	0.24	3.60	1.86	0.82	6.28
9.5mm thick to sides of openings less than 600mm wide	m	0.26	3.90	1.86	0.86	6.62
12.5mm thick to walls over 300mm wide	m2	0.42	6.30	5.24	1.73	13.27
12.5mm thick to columns over 600mm wide	m2	0.46	6.90	5.24	1.82	13.96
12.5mm thick to beams less than 600mm wide	m	0.26	3.90	2.23	0.92	7.05
12.5mm thick to sides of openings less than 600mm wide	m	0.28	4.20	2.23	0.96	7.39

	Unit	Labour Hours	Mat'ls £	O & P £	Total £

SUSPENDED CEILINGS

Gyproc M/F suspended ceiling
system complete with hangers
and channels, and filled in with

	Unit	Labour Hours	Mat'ls £	O & P £	Total £
single layer, tapered edge wallboard, 12.5mm thick	m2	0.80	12.00	13.02	3.75	28.77
double layer, tapered edge wallboard, 12.5mm thick	m2	0.95	14.25	18.01	4.84	37.10
single layer, tapered edge Fireline board, 12.5mm thick	m2	0.80	12.00	14.99	4.05	31.04
double layer, tapered edge Fireline board, 12.5mm thick	m2	0.95	14.25	21.21	5.32	40.78

Concealed 'Z' system suspended
ceiling complete with hangers
and channels, filled in with
single layer of acoustic tiles

size 600 x 600 x 19mm thick	m2	0.85	12.75	19.30	4.81	36.86

SCREEDS AND GRANOLITHIC

**Cement and sand (1:3) screed,
screeded finish**

Floors, level and to falls less
than 15° to the horizontal

25mm thick						
over 300mm wide	m2	0.30	4.50	3.26	1.16	8.92
less than 300mm wide	m	0.12	1.80	1.12	0.44	3.36

	Unit	Labour Hours	Mat'ls £	O & P £	Total £
32mm thick					
over 300mm wide	m2	0.32	4.80	3.27	1.21	9.28
less than 300mm wide	m	0.13	1.95	1.32	0.49	3.76
38mm thick					
over 300mm wide	m2	0.34	5.10	4.05	1.37	10.52
less than 300mm wide	m	0.14	2.10	1.99	0.61	4.70
50mm thick					
over 300mm wide	m2	0.36	5.40	4.98	1.56	11.94
less than 300mm wide	m	0.15	2.25	1.68	0.59	4.52
65mm thick					
over 300mm wide	m2	0.40	6.00	6.20	1.83	14.03
less than 300mm wide	m	0.16	2.40	2.22	0.69	5.31

Floors, level and to falls more than 15° to the horizontal

	Unit	Labour Hours	Mat'ls £	O & P £	Total £
25mm thick					
over 300mm wide	m2	0.34	5.10	3.26	1.25	9.61
less than 300mm wide	m	0.14	2.10	1.12	0.48	3.70
32mm thick					
over 300mm wide	m2	0.35	5.25	3.27	1.28	9.80
less than 300mm wide	m	0.16	2.40	1.32	0.56	4.28
38mm thick					
over 300mm wide	m2	0.40	6.00	4.05	1.51	11.56
less than 300mm wide	m	0.18	2.70	1.99	0.70	5.39
50mm thick					
over 300mm wide	m2	0.52	7.80	4.98	1.92	14.70
less than 300mm wide	m	0.20	3.00	1.68	0.70	5.38

	Unit	Labour Hours	Mat'ls £	O & P £	Total £

Cement and sand (1:3) screed,
(cont'd)

	Unit	Labour Hours	Mat'ls £	O & P £	Total £
65mm thick					
over 300mm wide	m2	0.56	8.40	6.20	2.19 16.79
less than 300mm wide	m	0.22	3.30	2.22	0.83 6.35

Linings to channels to slight
falls

	Unit	Labour Hours	Mat'ls £	O & P £	Total £
25mm thick					
150mm wide	m	0.15	2.25	0.51	0.41 3.17
250mm wide	m	0.20	3.00	0.91	0.59 4.50
400mm wide	m	0.25	3.75	1.44	0.78 5.97
32mm thick					
150mm wide	m	0.16	2.40	0.61	0.45 3.46
250mm wide	m	0.22	3.30	1.02	0.65 4.97
400mm wide	m	0.27	4.05	1.64	0.85 6.54
38mm thick					
150mm wide	m	0.17	2.55	0.66	0.48 3.69
250mm wide	m	0.23	3.45	1.02	0.67 5.14
400mm wide	m	0.28	4.20	1.73	0.89 6.82
50mm thick					
150mm wide	m	0.18	2.70	0.83	0.53 4.06
250mm wide	m	0.24	3.60	1.37	0.75 5.72
400mm wide	m	0.29	4.35	2.19	0.98 7.52
65mm thick					
150mm wide	m	0.20	3.00	1.06	0.61 4.67
250mm wide	m	0.26	3.90	1.77	0.85 6.52
400mm wide	m	0.32	4.80	2.70	1.13 8.63

	Unit	Labour Hours	Mat'ls £	O & P £	Total £

Landings

25mm thick
| over 300mm wide | m2 | 0.34 | 5.10 | 3.26 | 1.25 | 9.61 |
| less than 300mm wide | m | 0.11 | 1.65 | 1.12 | 0.42 | 3.19 |

32mm thick
| over 300mm wide | m2 | 0.36 | 5.40 | 3.72 | 1.37 | 10.49 |
| less than 300mm wide | m | 0.12 | 1.80 | 1.32 | 0.47 | 3.59 |

38mm thick
| over 300mm wide | m2 | 0.38 | 5.70 | 4.05 | 1.46 | 11.21 |
| less than 300mm wide | m | 0.13 | 1.95 | 1.99 | 0.59 | 4.53 |

50mm thick
| over 300mm wide | m2 | 0.40 | 6.00 | 4.98 | 1.65 | 12.63 |
| less than 300mm wide | m | 0.14 | 2.10 | 1.68 | 0.57 | 4.35 |

65mm thick
| over 300mm wide | m2 | 0.42 | 6.30 | 6.20 | 1.88 | 14.38 |
| less than 300mm wide | m | 0.15 | 2.25 | 2.22 | 0.67 | 5.14 |

Treads

25mm thick
| 250mm wide | m | 0.28 | 4.20 | 0.82 | 0.75 | 5.77 |
| 350mm wide | m | 0.30 | 4.50 | 1.11 | 0.84 | 6.45 |

32mm thick
| 250mm wide | m | 0.30 | 4.50 | 0.94 | 0.82 | 6.26 |
| 350mm wide | m | 0.32 | 4.80 | 1.13 | 0.89 | 6.82 |

38mm thick
| 250mm wide | m | 0.32 | 4.80 | 1.03 | 0.87 | 6.70 |
| 350mm wide | m | 0.34 | 5.10 | 1.40 | 0.98 | 7.48 |

	Unit	Labour	Hours £	Mat'ls £	O & P £	Total £
Cement and sand (1:3) screed, **(cont'd)**						
50mm thick						
250mm wide	m	0.34	5.10	1.25	0.95	7.30
350mm wide	m	0.36	5.40	1.77	1.08	8.25
65mm thick						
250mm wide	m	0.36	5.40	1.61	1.05	8.06
350mm wide	m	0.38	5.70	2.26	1.19	9.15
Risers						
13mm thick						
150mm high, plain	m	0.28	4.20	0.41	0.69	5.30
150mm high, undercut	m	0.32	4.80	0.41	0.78	5.99
19mm thick						
150mm high, plain	m	0.28	4.20	0.44	0.70	5.34
150mm high, undercut	m	0.32	4.80	0.44	0.79	6.03
200mm high, plain	m	0.30	4.50	0.47	0.75	5.72
200mm high, undercut	m	0.34	5.10	0.47	0.84	6.41
25mm thick						
150mm high, plain	m	0.34	5.10	0.48	0.84	6.42
150mm high, undercut	m	0.38	5.70	0.48	0.93	7.11
200mm high, plain	m	0.36	5.40	0.65	0.91	6.96
200mm high, undercut	m	0.40	6.00	0.65	1.00	7.65

	Unit	Labour	Hours £	Mat'ls £	O & P £	Total £
Surface dressings						
Liquid hardening agent	m2	0.05	0.75	0.64	0.21	1.60
Oil repellent agent	m2	0.05	0.75	0.31	0.16	1.22

Granolithic, cement and granite chippings steel trowelled smooth, to falls less than 15° to the horizontal

Floors, level and to falls less than 15° to the horizontal

	Unit	Labour	Hours £	Mat'ls £	O & P £	Total £
25mm thick						
over 300mm wide	m2	0.40	6.00	4.70	1.61	12.31
less than 300mm wide	m	0.14	2.10	1.47	0.54	4.11
32mm thick						
over 300mm wide	m2	0.42	6.30	5.13	1.71	13.14
less than 300mm wide	m	0.15	2.25	1.58	0.57	4.40
38mm thick						
over 300mm wide	m2	0.43	6.45	5.54	1.80	13.79
less than 300mm wide	m	0.16	2.40	1.72	0.62	4.74
50mm thick						
over 300mm wide	m2	0.46	6.90	6.37	1.99	15.26
less than 300mm wide	m	0.17	2.55	1.96	0.68	5.19
65mm thick						
over 300mm wide	m2	0.50	7.50	7.83	2.30	17.63
less than 300mm wide	m	0.18	2.70	2.40	0.77	5.87

	Unit	Labour Hours	Mat'ls £	O & P £	Total £
Floors, level and to falls more than 15° to the horizontal					
25mm thick					
over 300mm wide	m2	0.44	6.60	4.70	1.70	13.00
less than 300mm wide	m	0.16	2.40	1.47	0.58	4.45
32mm thick					
over 300mm thick	m2	0.46	6.90	5.13	1.80	13.83
less than 300mm thick	m	0.18	2.70	1.58	0.64	4.92
38mm thick					
over 300mm wide	m2	0.48	7.20	5.54	1.91	14.65
less than 300mm wide	m	0.20	3.00	1.72	0.71	5.43
50mm thick					
over 300mm wide	m2	0.52	7.80	6.37	2.13	16.30
less than 300mm wide	m	0.22	3.30	1.96	0.79	6.05
50mm thick					
over 300mm wide	m2	0.56	8.40	7.83	2.43	18.66
less than 300mm wide	m	0.24	3.60	2.40	0.90	6.90
Linings to channels to slight falls					
25mm thick					
150mm wide	m	0.17	2.55	0.70	0.49	3.74
250mm wide	m	0.22	3.30	1.18	0.67	5.15
400mm wide	m	0.27	4.05	1.91	0.89	6.85
32mm thick					
150mm wide	m	0.18	2.70	0.77	0.52	3.99
250mm wide	m	0.24	3.60	1.29	0.73	5.62
400mm wide	m	0.29	4.35	2.05	0.96	7.36

	Unit	Labour	Hours £	Mat'ls £	O & P £	Total £
38mm thick						
150mm wide	m	0.20	3.00	0.84	0.58	4.42
250mm wide	m	0.25	3.75	1.40	0.77	5.92
400mm wide	m	0.30	4.50	2.20	1.01	7.71
50mm thick						
150mm wide	m	0.22	3.30	0.96	0.64	4.90
250mm wide	m	0.27	4.05	1.60	0.85	6.50
400mm wide	m	0.32	4.80	2.57	1.11	8.48
65mm thick						
150mm wide	m	0.24	3.60	1.17	0.72	5.49
250mm wide	m	0.28	4.20	1.96	0.92	7.08
400mm wide	m	0.34	5.10	3.17	1.24	9.51

Landings

	Unit	Labour	Hours £	Mat'ls £	O & P £	Total £
25mm thick						
over 300mm wide	m2	0.44	6.60	4.70	1.70	13.00
less than 300mm wide	m	0.18	2.70	1.42	0.62	4.74
32mm thick						
over 300mm wide	m2	0.46	6.90	5.13	1.80	13.83
less than 300mm wide	m	0.20	3.00	1.58	0.69	5.27
38mm thick						
over 300mm wide	m2	0.48	7.20	5.54	1.91	14.65
less than 300mm wide	m	0.22	3.30	1.72	0.75	5.77
50mm thick						
over 300mm wide	m2	0.52	7.80	6.37	2.13	16.30
less than 300mm wide	m	0.25	3.75	1.96	0.86	6.57

	Unit	Labour Hours	Mat'ls £	O & P £	Total £
			£		
65mm thick					
over 300mm wide	m2	0.56	8.40	7.83	2.43	18.66
less than 300mm wide	m	0.28	4.20	2.40	0.99	7.59
Treads					
25mm thick					
250mm wide	m	0.32	4.80	1.16	0.89	6.85
350mm wide	m	0.34	5.10	1.67	1.02	7.79
32mm thick					
250mm wide	m	0.34	5.10	1.24	0.95	7.29
350mm wide	m	0.36	5.40	1.74	1.07	8.21
38mm thick					
250mm wide	m	0.36	5.40	1.33	1.01	7.74
350mm wide	m	0.38	5.70	1.82	1.13	8.65
50mm thick					
250mm wide	m	0.38	5.70	1.54	1.09	8.33
350mm wide	m	0.40	6.00	2.02	1.20	9.22
65mm thick					
250mm wide	m	0.42	6.30	1.88	1.23	9.41
350mm wide	m	0.42	6.30	2.49	1.32	10.11
Risers					
13mm thick					
150mm high, plain	m	0.28	4.20	0.39	0.69	5.28
150mm high, undercut	m	0.32	4.80	0.39	0.78	5.97

	Unit	Labour Hours	£	Mat'ls £	O & P £	Total £
19mm thick						
150mm high, plain	m	0.28	4.20	0.47	0.70	5.37
150mm high, undercut	m	0.32	4.80	0.47	0.79	6.06
200mm high, plain	m	0.30	4.50	0.63	0.67	5.80
200mm high, undercut	m	0.34	5.10	0.63	0.86	6.59
25mm thick						
150mm high, plain	m	0.34	5.10	0.72	0.87	6.69
150mm high, undercut	m	0.38	5.70	0.72	0.96	7.38
200mm high, plain	m	0.36	5.40	0.94	0.95	7.29
200mm high, undercut	m	0.40	6.00	0.94	1.04	7.98

PLASTERED COATINGS

Backing coats

Cement and sand (1:3) backing
coat to brickwork or blockwork
walls

	Unit	Labour Hours	£	Mat'ls £	O & P £	Total £
13mm thick						
over 300mm wide	m2	0.50	7.50	1.37	1.33	10.20
less than 300mm wide	m	0.20	3.00	0.49	0.52	4.01
19mm thick						
over 300mm wide	m2	0.58	8.70	2.47	1.68	12.85
less than 300mm wide	m	0.22	3.30	0.88	0.63	4.81

Cement and sand (1:3) backing
coat to brickwork or blockwork
isolated columns

	Unit	Labour Hours	£	Mat'ls £	O & P £	Total £
13mm thick						
over 300mm wide	m2	0.60	9.00	1.37	1.56	11.93
less than 300mm wide	m	0.24	3.60	0.49	0.61	4.70

	Unit	Labour Hours	Mat'ls £	O & P £	Total £

Cement and sand (1:3) backing
coat to brickwork or blockwork
isolated columns (cont'd)

19mm thick

	Unit	Labour Hours	Mat'ls £	O & P £	Total £
over 300mm wide	m2	0.68	10.20	2.47	1.90	14.57
less than 300mm wide	m	0.26	3.90	0.88	0.72	5.50

Cement and sand (1:3) backing
coat to brickwork or blockwork
isolated beams

13mm thick

over 300mm wide	m2	0.72	10.80	1.37	1.83	14.00
less than 300mm wide	m	0.30	4.50	0.49	0.75	5.74

19mm thick

over 300mm wide	m2	0.80	12.00	2.47	2.17	16.64
less than 300mm wide	m	0.32	4.80	0.88	0.85	6.53

Plasterwork

Pre-mixed lightweight plaster,
11mm bonding, 2mm finish

brickwork or blockwork walls

over 300mm wide	m2	0.48	7.20	2.66	1.48	11.34
less than 300mm wide	m	0.18	2.70	0.92	0.54	4.16

isolated columns

over 300mm wide	m2	0.58	8.70	2.66	1.70	13.06
less than 300mm wide	m	0.22	3.30	0.92	0.63	4.85

isolated beams

over 300mm wide	m2	0.70	10.50	2.66	1.97	15.13
less than 300mm wide	m	0.24	3.60	0.92	0.68	5.20

	Unit	Labour	Hours £	Mat'ls £	O & P £	Total £
ceilings						
over 300mm wide	m2	0.62	9.30	2.66	1.79	13.75
less than 300mm wide	m	0.24	3.60	0.92	0.68	5.20

One coat 'Universal' plaster to
brickwork or blockwork walls
13mm thick

brickwork or blockwork walls						
over 300mm wide	m2	0.30	4.50	3.31	1.17	8.98
less than 300mm wide	m	0.11	1.65	1.09	0.41	3.15
isolated columns						
over 300mm wide	m2	0.40	6.00	3.31	1.40	10.71
less than 300mm wide	m	0.14	2.10	1.09	0.48	3.67
isolated beams						
over 300mm wide	m2	0.50	7.50	3.31	1.62	12.43
less than 300mm wide	m	0.16	2.40	1.09	0.52	4.01
ceilings						
over 300mm wide	m2	0.34	5.10	3.31	1.26	9.67
less than 300mm wide	m	0.14	2.10	1.09	0.48	3.67

One coat 'Universal' plaster to
brickwork or blockwork walls
19mm thick

brickwork or blockwork walls						
over 300mm wide	m2	0.35	5.25	4.21	1.42	10.88
less than 300mm wide	m	0.13	1.95	0.45	0.36	2.76
isolated columns						
over 300mm wide	m2	0.45	6.75	4.21	1.64	12.60
less than 300mm wide	m	0.16	2.40	0.45	0.43	3.28

	Unit	Labour Hours	Mat'ls £	O & P £	Total £

One coat 'Universal' plaster to brickwork or blockwork walls 19mm thick (cont'd)

isolated beams

	Unit	Labour	Hours	Mat'ls	O & P	Total
over 300mm wide	m2	0.55	8.25	4.21	1.87	14.33
less than 300mm wide	m	0.18	2.70	0.45	0.47	3.62

ceilings

over 300mm wide	m2	0.40	6.00	4.21	1.53	11.74
less than 300mm wide	m	0.16	2.40	0.45	0.43	3.28

Two coat renovating plaster 11mm undercoat and 2mm finish

walls

over 300mm wide	m2	0.48	7.20	2.91	1.52	11.63
less than 300mm wide	m	0.18	2.70	0.98	0.55	4.23

isolated columns

over 300mm wide	m2	0.58	8.70	2.91	1.74	13.35
less than 300mm wide	m	0.22	3.30	0.98	0.64	4.92

isolated beams

over 300mm wide	m2	0.70	10.50	2.91	2.01	15.42
less than 300mm wide	m	0.24	3.60	0.98	0.69	5.27

ceilings

over 300mm wide	m2	0.62	9.30	2.91	1.83	14.04
less than 300mm wide	m	0.24	3.60	0.98	0.69	5.27

One coat board finish, 3mm thick

plasterboard walls

over 300mm wide	m2	0.26	3.90	1.60	0.83	6.33
less than 300mm wide	m	0.10	1.50	0.60	0.32	2.42

	Unit	Labour Hours	Mat'ls £	O & P £	Total £

plasterboard isolated columns

	Unit	Labour Hours	Mat'ls £	O & P £	Total £
over 300mm wide	m2	0.44	6.60	1.60	1.23	9.43
less than 300mm wide	m	0.16	2.40	0.60	0.45	3.45

plasterboard beams

over 300mm wide	m2	0.50	7.50	1.60	1.37	10.47
less than 300mm wide	m	0.18	2.70	0.60	0.50	3.80

plasterboard ceilings

over 300mm wide	m2	0.36	5.40	1.60	1.05	8.05
less than 300mm wide	m	0.14	2.10	0.60	0.41	3.11

METAL LATHING AND BEADS

Lathing

Expamet expanded metal lathing,
6mm, ref. BB263, 0.50mm thick

stapled to softwood, vertically

over 300mm wide	m2	0.26	3.90	4.77	1.30	9.97
less than 300mm wide	m	0.09	1.35	1.52	0.43	3.30

stapled to softwood, horizontally

over 300mm wide	m2	0.29	4.35	4.77	1.37	10.49
less than 300mm wide	m	0.11	1.65	1.52	0.48	3.65

tied with wire, vertically

over 300mm thick	m2	0.29	4.35	4.77	1.37	10.49
less than 300mm thick	m	0.11	1.65	1.52	0.48	3.65

tied with wire, horizontally

over 300mm wide	m2	0.32	4.80	4.77	1.44	11.01
less than 300mm wide	m	0.12	1.80	1.52	0.50	3.82

	Unit	Labour	Hours £	Mat'ls £	O & P £	Total £
Expamet expanded metal lathing, 6mm, ref. BB264, 0.725mm thick						
stapled to softwood, vertically						
over 300mm wide	m2	0.28	4.20	5.44	1.45	11.09
less than 300mm wide	m	0.10	1.50	1.54	0.46	3.50
stapled to softwood, horizontally						
over 300mm wide	m2	0.31	4.65	5.44	1.51	11.60
less than 300mm wide	m	0.12	1.80	1.54	0.50	3.84
tied with wire, vertically						
over 300mm wide	m2	0.31	4.65	5.44	1.51	11.60
less than 300mm wide	m	0.12	1.80	1.54	0.50	3.84
tied with wire, horizontally						
over 300mm wide	m2	0.34	5.10	5.44	1.58	12.12
less than 300mm wide	m	0.13	1.95	1.54	0.52	4.01
Expamet Riblath expanded metal lathing, ref. 269, 0.30mm thick						
stapled to softwood, vertically						
over 300mm wide	m2	0.28	4.20	5.96	1.52	11.68
less than 300mm wide	m	0.10	1.50	2.02	0.53	4.05
stapled to softwood, horizontally						
over 300mm wide	m2	0.31	4.65	5.96	1.59	12.20
less than 300mm wide	m	0.12	1.80	2.02	0.57	4.39
tied with wire, vertically						
over 300mm wide	m2	0.31	4.65	5.96	1.59	12.20
less than 300mm wide	m	0.12	1.80	2.02	0.57	4.39

	Unit	Labour Hours	Mat'ls £	O & P £	Total £
tied with wire, horizontally					
over 300mm wide	m2	0.34	5.10	5.96	1.66	12.72
less than 300mm wide	m	0.13	1.95	2.02	0.60	4.57

Expamet Riblath expanded metal lathing, ref. 271, 0.50mm thick

	Unit	Labour Hours	Mat'ls £	O & P £	Total £
stapled to softwood, vertically					
over 300mm wide	m2	0.28	4.20	6.91	1.67	12.78
less than 300mm wide	m	0.10	1.50	2.30	0.57	4.37
stapled to softwood, horizontally					
over 300mm wide	m2	0.31	4.65	6.91	1.73	13.29
less than 300mm wide	m	0.12	1.80	2.30	0.62	4.72
tied with wire, vertically					
over 300mm wide	m2	0.31	4.65	6.91	1.73	13.29
less than 300mm wide	m	0.12	1.80	2.30	0.62	4.72
tied with wire, horizontally					
over 300mm wide	m2	0.34	5.10	6.91	1.80	13.81
less than 300mm wide	m	0.13	1.95	2.30	0.64	4.89

Expamet Spraylath expanded metal lathing, ref. 273, 0.50mm thick

	Unit	Labour Hours	Mat'ls £	O & P £	Total £
stapled to softwood, vertically					
over 300mm wide	m2	0.28	4.20	8.45	1.90	14.55
less than 300mm wide	m	0.10	1.50	2.76	0.64	4.90
stapled to softwood, horizontally					
over 300mm wide	m2	0.31	4.65	8.45	1.97	15.07
less than 300mm wide	m	0.12	1.80	2.76	0.68	5.24

	Unit	Labour	Hours £	Mat'ls £	O & P £	Total £
tied with wire, vertically						
over 300mm wide	m2	0.31	4.65	8.45	1.97	15.07
less than 300mm wide	m	0.12	1.80	2.76	0.68	5.24
tied with wire, horizontally						
over 300mm wide	m2	0.34	5.10	8.45	2.03	15.58
less than 300mm wide	m	0.13	1.95	2.76	0.71	5.42

Beads

Expamet plaster beads fixed with plaster dabs

	Unit	Labour	Hours £	Mat'ls £	O & P £	Total £
stainless steel						
stop bead, ref. 546	m	0.10	1.50	0.96	0.37	2.83
stop bead, ref. 547	m	0.10	1.50	2.82	0.65	4.97
galvanised steel						
angle bead, ref. 550	m	0.10	1.50	0.96	0.37	2.83
architrave bead, ref. 579	m	0.10	1.50	1.52	0.45	3.47
stop bead, ref. 563	m	0.10	1.50	1.11	0.39	3.00

Expamet dry wall beads, nailed to softwood

	Unit	Labour	Hours £	Mat'ls £	O & P £	Total £
corner bead, ref. 548	m	0.10	1.50	1.10	0.39	2.99
edging bead, ref. 567	m	0.10	1.50	2.46	0.59	4.55

Expamet thin coat beads, nailed to softwood

	Unit	Labour	Hours £	Mat'ls £	O & P £	Total £
angle bead, ref. 553	m	0.10	1.50	1.21	0.41	3.12
stop bead, ref. 560	m	0.10	1.50	1.47	0.45	3.42

	Unit	Labour Hours	Mat'ls £	O & P £	Total £

FLOOR COVERINGS

Quarry tiling

Red quarry tiles, bedded, jointed
and pointed in cement mortar (1:3),
butt jointed, straight both ways,
to floors

	Unit	Labour Hours	Mat'ls £	O & P £	Total £
150 × 150 × 12.5mm thick					
over 300mm wide	m2	0.90	13.50	32.07	6.84	52.41
less than 300mm wide	m	0.36	5.40	11.24	2.50	19.14
200 × 200 × 19mm thick					
over 300mm wide	m2	0.80	12.00	52.71	9.71	74.42
less than 300mm wide	m	0.32	4.80	18.03	3.42	26.25

Red quarry tiles, bedded, jointed
and pointed in cement mortar (1:3),
butt jointed, straight both ways,
to landings

	Unit	Labour Hours	Mat'ls £	O & P £	Total £
150 × 150 × 12.5mm thick					
over 300mm wide	m2	1.00	15.00	32.07	7.06	54.13
less than 300mm wide	m	0.50	7.50	11.24	2.81	21.55
200 × 200 × 19mm thick					
over 300mm wide	m2	0.90	13.50	52.71	9.93	76.14
less than 300mm wide	m	0.36	5.40	18.03	3.51	26.94

	Unit	Labour Hours	Mat'ls £	O & P £	Total £
Red quarry tiles to cills, 150mm wide, with one rounded edge	m	0.38	5.70	9.17	2.23	17.10
Red quarry tiles to coved skirtings, 150mm high	m	0.32	4.80	9.87	2.20	16.87

	Unit	Labour £	Hours £	Mat'ls £	O & P £	Total £
Extra for						
raking cutting	m	0.10	1.50	-	0.23	1.73
curved cutting	m	0.12	1.80	-	0.27	2.07

Brown quarry tiles, bedded, jointed
and pointed in cement mortar (1:3),
butt jointed, straight both ways,
to floors

	Unit	Labour £	Hours £	Mat'ls £	O & P £	Total £
150 × 150 × 12.5mm thick						
over 300mm wide	m2	0.90	13.50	34.60	7.22	55.32
less than 300mm wide	m	0.36	5.40	11.98	2.61	19.99
200 × 200 × 19mm thick						
over 300mm wide	m2	0.80	12.00	55.47	10.12	77.59
less than 300mm wide	m	0.32	4.80	19.03	3.57	27.40

Brown quarry tiles, bedded, jointed
and pointed in cement mortar (1:3),
butt jointed, straight both ways,
to landings

	Unit	Labour £	Hours £	Mat'ls £	O & P £	Total £
150 × 150 × 12.5mm thick						
over 300mm wide	m2	1.00	15.00	34.60	7.44	57.04
less than 300mm wide	m	0.50	7.50	11.98	2.92	22.40
200 × 200 × 19mm thick						
over 300mm wide	m2	0.80	12.00	55.47	10.12	77.59
less than 300mm wide	m	0.32	4.80	19.03	3.57	27.40
Brown quarry tiles to cills, 150mm wide, with one rounded edge	m	0.38	5.70	9.60	2.30	17.60
Brown quarry tiles to coved skirtings, 150mm high	m	0.32	4.80	10.35	2.27	17.42

	Unit	Labour	Hours £	Mat'ls £	O & P £	Total £
Extra for						
raking cutting	m	0.10	1.50	-	0.23	1.73
curved cutting	m	0.12	1.80	-	0.27	2.07

Ceramic tiling

Vitrified ceramic tiles, bedded,
jointed and pointed in cement
mortar (1:3), butt jointed, straight
both ways, to floors

	Unit	Labour	Hours £	Mat'ls £	O & P £	Total £
100 × 100 × 9mm thick						
over 300mm wide	m2	1.00	15.00	26.96	6.29	48.25
less than 300mm wide	m	0.44	6.60	9.82	2.46	18.88
150 × 150 × 12.5mm thick						
over 300mm wide	m2	0.90	13.50	24.38	5.68	43.56
less than 300mm wide	m	0.36	5.40	8.19	2.04	15.63
200 × 200 × 12.5mm thick						
over 300mm wide	m2	0.80	12.00	21.32	5.00	38.32
less than 300mm wide	m	0.32	4.80	6.89	1.75	13.44

Vitrified ceramic tiles, bedded,
jointed and pointed in cement
mortar (1:3), butt jointed, straight
both ways, to landings

	Unit	Labour	Hours £	Mat'ls £	O & P £	Total £
100 × 100 × 9mm thick						
over 300mm wide	m2	1.10	16.50	26.96	6.52	49.98
less than 300mm wide	m	0.54	8.10	9.82	2.69	20.61
150 × 150 × 12.5mm thick						
over 300mm wide	m2	1.00	15.00	24.38	5.91	45.29
less than 300mm wide	m	0.44	6.60	8.19	2.22	17.01

	Unit	Labour Hours	£	Mat'ls £	O & P £	Total £
200 × 200 × 12.5mm thick						
over 300mm wide	m2	0.98	14.70	21.32	5.40	41.42
less than 300mm wide	m	0.42	6.30	6.89	1.98	15.17
Extra for						
raking cutting	m	0.08	1.20	-	0.18	1.38
curved cutting	m	0.10	1.50	-	0.23	1.73

Terrazzo tiling

Terrazzo tile paving, bedded and
pointed in white cement mortar
(1:3) 12mm thick, butt jointed
straight both ways, to floors

	Unit	Labour Hours	£	Mat'ls £	O & P £	Total £
300 × 300 × 28mm thick						
over 300mm wide	m2	1.60	24.00	18.90	6.44	49.34
less than 300mm wide	m	0.65	9.75	9.15	2.84	21.74

Terrazzo tile paving, bedded and
pointed in white cement mortar
(1:3) 12mm thick, butt jointed
straight both ways, to landings

	Unit	Labour Hours	£	Mat'ls £	O & P £	Total £
300 × 300 × 28mm thick						
over 300mm wide	m2	1.70	25.50	18.90	6.66	51.06
less than 300mm wide	m	0.70	10.50	9.15	2.95	22.60

Vinyl tiling

Vinyl floor tiling, fixed with
adhesive, butt jointed straight
both ways, to floors

	Unit	Labour	Hours £	Mat'ls £	O & P £	Total £
300 × 300 × 2mm thick						
over 300mm wide	m2	0.28	4.20	8.34	1.88	14.42
less than 300mm wide	m	0.10	1.50	2.56	0.61	4.67

Vinyl floor tiling, fixed with
adhesive, butt jointed straight
both ways, to landings

	Unit	Labour	Hours £	Mat'ls £	O & P £	Total £
300 × 300 × 2mm thick						
over 300mm wide	m2	0.36	5.40	8.34	2.06	15.80
less than 300mm wide	m	0.12	1.80	2.56	0.65	5.01

Vinyl floor tiling, fixed with
adhesive, butt jointed straight
both ways, to floors

	Unit	Labour	Hours £	Mat'ls £	O & P £	Total £
300 × 300 × 2.5mm thick						
over 300mm wide	m2	0.28	4.20	8.62	1.92	14.74
less than 300mm wide	m	0.10	1.50	2.58	0.61	4.69

Vinyl floor tiling, fixed with
adhesive, butt jointed straight
both ways, to landings

	Unit	Labour	Hours £	Mat'ls £	O & P £	Total £
300 × 300 × 2.5mm thick						
over 300mm wide	m2	0.36	5.40	8.62	2.10	16.12
less than 300mm wide	m	0.12	1.80	2.58	0.66	5.04

Cork tiling

Cork floor tiling, fixed with
adhesive, butt jointed straight
both ways, to floors

	Unit	Labour	Hours £	Mat'ls £	O & P £	Total £

Cork tiling (cont'd)

300 × 300 × 4mm thick

	Unit	Labour	Hours £	Mat'ls £	O & P £	Total £
over 300mm wide	m2	0.36	5.40	24.92	4.55	34.87
less than 300mm wide	m	0.12	1.80	8.65	1.57	12.02

Cork floor tiling, fixed with
adhesive, butt jointed straight
both ways, to landings

300 × 300 × 4mm thick

	Unit	Labour	Hours £	Mat'ls £	O & P £	Total £
over 300mm wide	m2	0.42	6.30	11.51	2.67	20.48
less than 300mm wide	m	0.14	2.10	3.96	0.91	6.97

Carpet tiling

Carpet floor tiling, fixed with
adhesive, butt jointed straight
both ways, to floors

500 × 500

	Unit	Labour	Hours £	Mat'ls £	O & P £	Total £
over 300mm wide	m2	0.28	4.20	11.51	2.36	18.07
less than 300mm wide	m	0.10	1.50	3.96	0.82	6.28

Carpet floor tiling, fixed with
adhesive, butt jointed straight
both ways, to landings

300 × 300

	Unit	Labour	Hours £	Mat'ls £	O & P £	Total £
over 300mm wide	m2	0.42	6.30	11.51	2.67	20.48
less than 300mm wide	m	0.14	2.10	3.96	0.91	6.97

	Unit	Labour Hours	Mat'ls £	O & P £	Total £

Vinyl sheeting

Vinyl floor sheeting with welded
joints, fixed with adhesive

2mm thick to floors
over 300mm wide	m2	0.36	5.40	11.18	2.49	19.07
less than 300mm wide	m	0.14	2.10	4.19	0.94	7.23

2mm thick to landings
over 300mm wide	m2	0.40	6.00	11.18	2.58	19.76
less than 300mm wide	m	0.16	2.40	4.19	0.99	7.58

2.5mm thick to floors
over 300mm thick	m2	0.40	6.00	14.92	3.14	24.06
less than 300mm thick	m	0.18	2.70	5.21	1.19	9.10

2.5mm thick to landings
over 300mm wide	m2	0.44	6.60	14.92	3.23	24.75
less than 300mm wide	m	0.20	3.00	5.21	1.23	9.44

Linoleum

Linoleum floor sheeting with
welded joints, fixed with adhesive

2mm thick
over 300mm wide	m2	0.36	5.40	11.57	2.55	19.52
less than 300mm wide	m	0.14	2.10	4.26	0.95	7.31

2.5mm thick
over 300mm wide	m2	0.40	6.00	12.59	2.79	21.38
less than 300mm wide	m	0.18	2.70	5.27	1.20	9.17

	Unit	Labour £	Hours £	Mat'ls £	O & P £	Total £

Woodblock flooring

Woodblock flooring, 25mm thick,
tongued and grooved, laid herring-
bone pattern, fixing with adhesive,
to floors

maple
| over 300mm wide | m2 | 1.10 | 16.50 | 50.82 | 10.10 | 77.42 |
| less than 300mm wide | m | 0.36 | 5.40 | 19.71 | 3.77 | 28.88 |

european oak
| over 300mm wide | m2 | 1.10 | 16.50 | 44.78 | 9.19 | 70.47 |
| less than 300mm wide | m | 0.36 | 5.40 | 17.81 | 3.48 | 26.69 |

iroko
| over 300mm wide | m2 | 1.10 | 16.50 | 46.60 | 9.47 | 72.57 |
| less than 300mm wide | m | 0.36 | 5.40 | 18.57 | 3.60 | 27.57 |

merbau
| over 300mm wide | m2 | 1.10 | 16.50 | 46.60 | 9.47 | 72.57 |
| less than 300mm wide | m | 0.36 | 5.40 | 18.57 | 3.60 | 27.57 |

sapele
| over 300mm wide | m2 | 1.10 | 16.50 | 40.68 | 8.58 | 65.76 |
| less than 300mm wide | m | 0.36 | 5.40 | 16.71 | 3.32 | 25.43 |

Woodblock flooring, 25mm thick,
tongued and grooved, laid herring-
bone pattern, fixing with adhesive,
to landings

maple
| over 300mm wide | m2 | 1.20 | 18.00 | 50.82 | 10.32 | 79.14 |
| less than 300mm wide | m | 0.40 | 6.00 | 19.71 | 3.86 | 29.57 |

	Unit	Labour	Hours £	Mat'ls £	O & P £	Total £
european oak						
over 300mm wide	m2	1.20	18.00	44.78	9.42	72.20
less than 300mm wide	m	0.40	6.00	17.81	3.57	27.38
iroko						
over 300mm wide	m2	1.20	18.00	46.60	9.69	74.29
less than 300mm wide	m	0.40	6.00	18.57	3.69	28.26
merbau						
over 300mm thick	m2	1.20	18.00	42.88	9.13	70.01
less than 300mm thick	m	0.40	6.00	17.20	3.48	26.68
sapele						
over 300mm wide	m2	1.20	18.00	40.68	8.80	67.48
less than 300mm wide	m	0.40	6.00	16.71	3.41	26.12

WALL TILING

White glazed ceramic wall tiling
fixed with adhesive, pointing
with white grout

	Unit	Labour	Hours £	Mat'ls £	O & P £	Total £
100 × 100 × 4mm thick						
over 300mm wide	m2	0.95	14.25	27.60	6.28	48.13
less than 300mm wide	m	0.38	5.70	9.87	2.34	17.91
150 × 150 × 6.5mm thick						
over 300mm wide	m2	0.80	12.00	15.62	4.14	31.76
less than 300mm wide	m	0.35	5.25	6.21	1.72	13.18
200 × 200 × 6.5mm thick						
over 300mm wide	m2	0.75	11.25	16.78	4.20	32.23
less than 300mm wide	m	0.32	4.80	7.80	1.89	14.49

	Unit	Labour Hours	Mat'ls £	O & P £	Total £

White glazed ceramic wall tiling (cont'd)

200 × 250 × 6.5mm thick

over 300mm wide	m2	0.70	10.50	18.01	4.28	32.79
less than 300mm wide	m	0.30	4.50	7.17	1.75	13.42

250 × 265 × 6.5mm thick

over 300mm wide	m2	0.70	10.50	21.15	4.75	36.40
less than 300mm wide	m	0.30	4.50	7.41	1.79	13.70

250 × 330 × 6.5mm thick

over 300mm wide	m2	0.65	9.75	17.69	4.12	31.56
less than 300mm wide	m	0.28	4.20	7.32	1.73	13.25

250 × 330 × 6.5mm thick

over 300mm wide	m2	0.65	9.75	17.48	4.08	31.31
less than 300mm wide	m	0.28	4.20	7.23	1.71	13.14

250 × 400 × 6.5mm thick

over 300mm wide	m2	0.65	9.75	23.23	4.95	37.93
less than 300mm wide	m	0.28	4.20	7.91	1.82	13.93

310 × 450 × 6.5mm thick

over 300mm wide	m2	0.60	9.00	24.69	5.05	38.74
less than 300mm wide	m	0.28	4.20	8.62	1.92	14.74

330 × 400 × 6.5mm thick

over 300mm wide	m2	0.60	9.00	28.13	5.57	42.70
less than 300mm wide	m	0.26	3.90	9.90	2.07	15.87

To cills, with one rounded edge

100mm wide	m	0.24	3.60	2.39	0.90	6.89
150mm wide	m	0.30	4.50	3.58	1.21	9.29

	Unit	Labour Hours £	Mat'ls £	O & P £	Total £
Extra for					
raking cutting	m	0.10	1.50	-	0.23	1.73
curved cutting	m	0.15	2.25	-	0.34	2.59
holes for small pipe	nr	0.20	3.00	-	0.45	3.45
holes for large pipe	nr	0.25	3.75	-	0.56	4.31

Note: The table values are: raking cutting m 0.10 1.50 - 0.23 1.73

Coloured glazed ceramic wall tiling fixed with adhesive, pointing with white grout

	Unit	Labour Hours £	Mat'ls £	O & P £	Total £
100 × 100 × 4mm thick					
over 300mm wide	m2	0.95	14.25	29.78	6.60	50.63
less than 300mm wide	m	0.38	5.70	10.48	2.43	18.61
150 × 150 × 6.5mm thick					
over 300mm wide	m2	0.85	12.75	17.10	4.48	34.33
less than 300mm wide	m	0.35	5.25	7.18	1.86	14.29
200 × 200 × 6.5mm thick					
over 300mm wide	m2	0.75	11.25	18.02	4.39	33.66
less than 300mm wide	m	0.32	4.80	7.42	1.83	14.05
200 × 250 × 6.5mm thick					
over 300mm wide	m2	0.70	10.50	19.41	4.49	34.40
less than 300mm wide	m	0.30	4.50	7.58	1.81	13.89
250 × 265 × 6.5mm thick					
over 300mm wide	m2	0.70	10.50	22.40	4.94	37.84
less than 300mm wide	m	0.30	4.50	8.21	1.91	14.62
250 × 330 × 6.5mm thick					
over 300mm wide	m2	0.65	9.75	19.19	4.34	33.28
less than 300mm wide	m	0.28	4.20	7.47	1.75	13.42

	Unit	Labour Hours	Mat'ls £	O & P £	Total £
250 × 330 × 6.5mm thick					
over 300mm wide	m2	0.65	9.75	19.47	4.38	33.60
less than 300mm wide	m	0.28	4.20	7.14	1.70	13.04
250 × 400 × 6.5mm thick					
over 300mm wide	m2	0.65	9.75	24.92	5.20	39.87
less than 300mm wide	m	0.32	4.80	9.27	2.11	16.18
310 × 450 × 6.5mm thick					
over 300mm wide	m2	0.60	9.00	26.10	5.27	40.37
less than 300mm wide	m	0.28	4.20	10.10	2.15	16.45
330 × 400 × 6.5mm thick					
over 300mm wide	m2	0.60	9.00	29.97	5.85	44.82
less than 300mm wide	m	0.26	3.90	10.58	2.17	16.65
To cills, with one rounded edge					
100mm wide	m	0.24	3.60	3.18	1.02	7.80
150mm wide	m	0.30	4.50	4.62	1.37	10.49
Extra for					
raking cutting	m	0.10	1.50	-	0.23	1.73
curved cutting	m	0.12	1.80	-	0.27	2.07
holes for small pipe	nr	0.20	3.00	-	0.45	3.45
holes for large pipe	nr	0.25	3.75	-	0.56	4.31

	Unit	Labour Hours £	Mat'ls £	O & P £	Total £

PARTITIONS AND DRY LININGS

Laminated partitions

Gyproc laminated partition,
consisting of two skins of 12.5mm
thick tapered edge wallboard
bonded to 19mm thick
plasterboard core, joints taped
and filled, one coat drywall top
coat

	Unit	Labour Hours	Mat'ls	O & P	Total
50mm thick					
over 300mm wide	m2	1.00	15.00	13.00	4.20	32.20
less than 300mm wide	m	0.40	6.00	5.10	1.67	12.77

Fixing partition to other surfaces
including 38 x 25mm batten

to base	m	0.16	2.40	0.66	0.46	3.52
to head	m	0.20	3.00	0.66	0.55	4.21
to side	m	0.18	2.70	0.66	0.50	3.86
Raking cutting	m	0.14	2.10	-	0.32	2.42

Cutting around steel angles, joists,
pipes and trunking

over 2m girth	nr	0.14	2.10	-	0.32	2.42
less than 300mm girth	nr	0.06	0.90	-	0.14	1.04
300mm to 1m girth	nr	0.08	1.20	-	0.18	1.38
1m to 2m girth	nr	0.10	1.50	-	0.23	1.73

	Unit	Labour Hours	Mat'ls £	O & P £	Total £

Gyproc laminated partition,
consisting of two skins of 12.5mm
thick tapered edge wallboard
bonded to 19mm thick
plasterboard core, joints taped
and filled, one coat drywall top
coat

	Unit	Labour Hours	Mat'ls £	O & P £	Total £
65mm thick					
over 300mm wide	m2	1.10	16.50	16.12	4.89	37.51
less than 300mm wide	m	0.45	6.75	6.13	1.93	14.81

Fixing partition to other surfaces
including 38 x 25mm batten

	Unit	Labour Hours	Mat'ls £	O & P £	Total £
to base	m	0.16	2.40	0.66	0.46	3.52
to head	m	0.20	3.00	0.66	0.55	4.21
to side	m	0.18	2.70	0.66	0.50	3.86

	Unit	Labour Hours	Mat'ls £	O & P £	Total £
Raking cutting	m	0.14	2.10	-	0.32	2.42

Cutting around steel angles, joists,
pipes and trunking

	Unit	Labour Hours	Mat'ls £	O & P £	Total £
over 2m girth	nr	0.14	2.10	-	0.32	2.42
less than 300mm girth	nr	0.06	0.90	-	0.14	1.04
300mm to 1m girth	nr	0.08	1.20	-	0.18	1.38
1m to 2m girth	nr	0.10	1.50	-	0.23	1.73

	Unit	Labour Hours £	Mat'ls £	O & P £	Total £

Metal stud partitions

Gyproc metal stud partition
consisting of metal stud framing
48mm wide, floor channel plugged
amd screwed to concrete floor,
two skins of 12.5mm thick
tapered edge wallboard bonded
to 19mm thick plasterboard
core, joints taped and filled,
one coat drywall top coat

	Unit	Labour Hours £	Mat'ls £	O & P £	Total £
75mm thick					
over 300mm wide	m2	1.05	15.75	10.15	3.89	29.79
less than 300mm wide	m	0.45	6.75	4.12	1.63	12.50

Fixing partition to other surfaces
including 38 x 25mm batten

to base	m	0.26	3.90	0.52	0.66	5.08
to head	m	0.30	4.50	0.52	0.75	5.77
to side	m	0.38	5.70	0.52	0.93	7.15
Raking cutting	m	0.24	3.60	-	0.54	4.14

Cutting around steel angles, joists,
pipes and trunking

over 2m girth	nr	0.24	3.60	-	0.54	4.14
less than 300mm girth	nr	0.16	2.40	-	0.36	2.76
300mm to 1m girth	nr	0.18	2.70	-	0.41	3.11
1m to 2m girth	nr	0.20	3.00	-	0.45	3.45

	Unit	Labour Hours £	Mat'ls £	O & P £	Total £

Gyproc metal stud partition
consisting of metal stud framing
70mm wide, floor channel plugged
and screwed to concrete floor,
two skins of 12.5mm thick
tapered edge wallboard bonded
to 19mm thick plasterboard
core, joints taped and filled,
one coat drywall top coat

	Unit	Labour	Hours £	Mat'ls £	O & P £	Total £
97mm thick						
over 300mm wide	m2	1.05	15.75	14.62	4.56	34.93
less than 300mm wide	m	0.45	6.75	5.40	1.82	13.97
Fixing partition to other surfaces **including 38 x 25mm batten**						
to base	m	0.26	3.90	0.52	0.66	5.08
to head	m	0.30	4.50	0.52	0.75	5.77
to side	m	0.38	5.70	0.52	0.93	7.15
Raking cutting	m	0.34	5.10	-	0.77	5.87
Cutting around steel angles, joists, **pipes and trunking**						
over 2m girth	nr	0.24	3.60	-	0.54	4.14
less than 300mm girth	nr	0.16	2.40	-	0.36	2.76
300mm to 1m girth	nr	0.18	2.70	-	0.41	3.11
1m to 2m girth	nr	0.20	3.00	-	0.45	3.45

	Unit	Labour Hours £	Mat'ls £	O & P £	Total £

Dry linings

Gypsum plasterboard with tapered
edges, fixing with galvanised nails,
joints taped and filled to receive
direct decoration

	Unit	Labour Hours £	Mat'ls £	O & P £	Total £
9.5mm thick					
to walls					
over 300mm wide	m2	0.28	4.20	2.34	0.98	7.52
less than 300mm wide	m	0.16	2.40	1.30	0.56	4.26
to ceilings					
over 300mm wide	m2	0.30	4.50	2.34	1.03	7.87
less than 300mm wide	m	0.18	2.70	1.30	0.60	4.60
to beams					
over 300mm wide	m2	0.33	4.95	2.34	1.09	8.38
less than 300mm wide	m	0.20	3.00	1.30	0.65	4.95
to columns					
over 300mm wide	m2	0.33	4.95	2.34	1.09	8.38
less than 300mm wide	m	0.20	3.00	1.30	0.65	4.95
Raking cutting	m	0.20	3.00	-	0.45	3.45

Cutting around steel angles, joists,
pipes and trunking

	Unit	Labour Hours £	Mat'ls £	O & P £	Total £
over 2m girth	nr	0.24	3.60	-	0.54	4.14
less than 300mm girth	nr	0.16	2.40	-	0.36	2.76
300mm to 1m girth	nr	0.18	2.70	-	0.41	3.11
1m to 2m girth	nr	0.20	3.00	-	0.45	3.45

	Unit	Labour Hours	Mat'ls £	O & P £	Total £

Gypsum plasterboard with tapered
edges, fixing with galvanised nails,
joints taped and filled to receive
direct decoration

	Unit	Labour Hours	Mat'ls £	O & P £	Total £
12.5mm thick					
to walls					
over 300mm wide	m2	0.30	4.50	2.70	1.08	8.28
less than 300mm wide	m	0.18	2.70	1.42	0.62	4.74
to ceilings					
over 300mm wide	m2	0.35	5.25	2.70	1.19	9.14
less than 300mm wide	m	0.20	3.00	1.42	0.66	5.08
to beams					
over 300mm wide	m2	0.40	6.00	2.70	1.31	10.01
less than 300mm wide	m	0.20	3.00	1.42	0.66	5.08
to columns					
over 300mm wide	m2	0.40	6.00	2.70	1.31	10.01
less than 300mm wide	m	0.20	3.00	1.42	0.66	5.08
Raking cutting	m	0.20	3.00	-	0.45	3.45

Cutting around steel angles, joists,
pipes and trunking

	Unit	Labour Hours	Mat'ls £	O & P £	Total £
over 2m girth	nr	0.24	3.60	-	0.54	4.14
less than 300mm girth	nr	0.16	2.40	-	0.36	2.76
300mm to 1m girth	nr	0.18	2.70	-	0.41	3.11
1m to 2m girth	nr	0.20	3.00	-	0.45	3.45

	Unit	Labour Hours	Mat'ls £	O & P £	Total £

Gypsum plasterboard with tapered
edges, fixing with galvanised nails,
joints taped and filled including
one coat board finish, 3mm thick

9.5mm thick
  to walls

	Unit	Labour Hours	Mat'ls £	O & P £	Total £
over 300mm wide	m2	0.54	8.10	3.71	1.77	13.58
less than 300mm wide	m	0.26	3.90	1.88	0.87	6.65
to ceilings						
over 300mm wide	m2	0.66	9.90	3.71	2.04	15.65
less than 300mm wide	m	0.32	4.80	1.88	1.00	7.68
to beams						
over 300mm wide	m2	0.84	12.60	3.71	2.45	18.76
less than 300mm wide	m	0.38	5.70	1.88	1.14	8.72
to columns						
over 300mm wide	m2	0.78	11.70	3.71	2.31	17.72
less than 300mm wide	m	0.34	5.10	1.88	1.05	8.03
Raking cutting	m	0.20	3.00	-	0.45	3.45

Cutting around steel angles, joists,
pipes and trunking

	Unit	Labour Hours	Mat'ls £	O & P £	Total £
over 2m girth	nr	0.24	3.60	-	0.54	4.14
less than 300mm girth	nr	0.16	2.40	-	0.36	2.76
300mm to 1m girth	nr	0.18	2.70	-	0.41	3.11
1m to 2m girth	nr	0.20	3.00	-	0.45	3.45

	Unit	Labour Hours	Mat'ls £	O & P £	Total £

Gypsum plasterboard with tapered
edges, fixing with galvanised nails,
joints taped and filled including
one coat board finish, 3mm thick

	Unit	Labour Hours	Mat'ls £	O & P £	Total £
12.5mm thick					
to walls					
over 300mm wide	m2	0.66	9.90	4.14	2.11	16.15
less than 300mm wide	m	0.32	4.80	2.00	1.02	7.82
to ceilings					
over 300mm wide	m2	0.71	10.65	4.14	2.22	17.01
less than 300mm wide	m	0.34	5.10	2.00	1.07	8.17
to beams					
over 300mm wide	m2	0.74	11.10	4.14	2.29	17.53
less than 300mm wide	m	0.36	5.40	2.00	1.11	8.51
to columns					
over 300mm wide	m2	0.74	11.10	4.14	2.29	17.53
less than 300mm wide	m	0.36	5.40	2.00	1.11	8.51
Raking cutting	m	0.20	3.00	-	0.45	3.45
Cutting around steel angles, joists, pipes and trunking					
over 2m girth	nr	0.24	3.60	-	0.54	4.14
less than 300mm girth	nr	0.16	2.40	-	0.36	2.76
300mm to 1m girth	nr	0.18	2.70	-	0.41	3.11
1m to 2m girth	nr	0.20	3.00	-	0.45	3.45

	Unit	Labour Hours	Mat'ls £	O & P £	Total £
Gypsum thermal board with tapered edges, fixing with adhesive to smooth surfaces, joints taped to receive direct decoration					
30mm thick					
to walls					
over 300mm wide	m2	0.34	5.10	8.50	2.04	15.64
less than 300mm wide	m	0.14	2.10	2.90	0.75	5.75
to ceilings					
over 300mm wide	m2	0.40	6.00	8.50	2.18	16.68
less than 300mm wide	m	0.16	2.40	2.90	0.80	6.10
to beams					
over 300mm wide	m2	0.40	6.00	8.50	2.18	16.68
less than 300mm wide	m	0.16	2.40	2.90	0.80	6.10
to columns					
over 300mm wide	m2	0.40	6.00	8.50	2.18	16.68
less than 300mm wide	m	0.16	2.40	2.90	0.80	6.10
Raking cutting	m	0.25	3.75	-	0.56	4.31
Cutting around steel angles, joists, pipes and trunking					
over 2m girth	nr	0.30	4.50	-	0.68	5.18
less than 300mm girth	nr	0.20	3.00	-	0.45	3.45
300mm to 1m girth	nr	0.22	3.30	-	0.50	3.80
1m to 2m girth	nr	0.24	3.60	-	0.54	4.14

	Unit	Labour Hours	Mat'ls £	O & P £	Total £

Gypsum thermal board with tapered
edges, fixing with adhesive to
smooth surfaces, joints taped
to receive direct decoration

	Unit	Labour Hours	Mat'ls £	O & P £	Total £
40mm thick					
to walls					
over 300mm wide	m2	0.38	5.70	9.02	2.21	16.93
less than 300mm wide	m	0.18	2.70	3.04	0.86	6.60
to ceilings					
over 300mm wide	m2	0.44	6.60	9.02	2.34	17.96
less than 300mm wide	m	0.18	2.70	3.04	0.86	6.60
to beams					
over 300mm wide	m2	0.50	7.50	9.02	2.48	19.00
less than 300mm wide	m	0.18	2.70	3.04	0.86	6.60
to columns					
over 300mm wide	m2	0.38	5.70	9.02	2.21	16.93
less than 300mm wide	m	0.18	2.70	3.04	0.86	6.60
Raking cutting	m	0.28	4.20	-	0.63	4.83
Cutting around steel angles, joists, pipes and trunking					
over 2m girth	nr	0.32	4.80	-	0.72	5.52
less than 300mm girth	nr	0.22	3.30	-	0.50	3.80
300mm to 1m girth	nr	0.26	3.90	-	0.59	4.49
1m to 2m girth	nr	0.28	4.20	-	0.63	4.83

	Unit	Labour	Hours £	Mat'ls £	O & P £	Total £
Gypsum thermal board with tapered edges, fixing with adhesive to smooth surfaces, joints taped to receive direct decoration						
50mm thick						
to walls						
over 300mm wide	m2	0.42	6.30	9.48	2.37	18.15
less than 300mm wide	m	0.13	1.95	3.14	0.76	5.85
to ceilings						
over 300mm wide	m2	0.48	7.20	9.48	2.50	19.18
less than 300mm wide	m	0.20	3.00	3.14	0.92	7.06
to beams						
over 300mm wide	m2	0.48	7.20	9.48	2.50	19.18
less than 300mm wide	m	0.20	3.00	3.14	0.92	7.06
to columns						
over 300mm wide	m2	0.48	7.20	9.48	2.50	19.18
less than 300mm wide	m	0.20	3.00	3.14	0.92	7.06
Raking cutting	m	0.32	4.80	-	0.72	5.52
Cutting around steel angles, joists, pipes and trunking						
over 2m girth	nr	0.36	5.40	-	0.81	6.21
less than 300mm girth	nr	0.26	3.90	-	0.59	4.49
300mm to 1m girth	nr	0.28	4.20	-	0.63	4.83
1m to 2m girth	nr	0.32	4.80	-	0.72	5.52

	Unit	Labour Hours	£	Mat'ls £	O & P £	Total £

Gypsum thermal board with tapered
edges, fixing with adhesive to
smooth surfaces, joints taped
including one coat board
finish, 3mm thick

	Unit	Labour	Hours £	Mat'ls £	O & P £	Total £
30mm thick						
to walls						
over 300mm wide	m2	0.60	9.00	10.12	2.87	21.99
less than 300mm wide	m	0.22	3.30	3.68	1.05	8.03
to ceilings						
over 300mm wide	m2	0.66	9.90	10.12	3.00	23.02
less than 300mm wide	m	0.24	3.60	3.68	1.09	8.37
to beams						
over 300mm wide	m2	0.66	9.90	10.12	3.00	23.02
less than 300mm wide	m	0.24	3.60	3.68	1.09	8.37
to columns						
over 300mm wide	m2	0.66	9.90	10.12	3.00	23.02
less than 300mm wide	m	0.24	3.60	3.68	1.09	8.37
Raking cutting	m	0.25	3.75	-	0.56	4.31
Cutting around steel angles, joists, pipes and trunking						
over 2m girth	nr	0.30	4.50	-	0.68	5.18
less than 300mm girth	nr	0.20	3.00	-	0.45	3.45
300mm to 1m girth	nr	0.22	3.30	-	0.50	3.80
1m to 2m girth	nr	0.24	3.60	-	0.54	4.14

	Unit	Labour Hours	Mat'ls £	O & P £	Total £

Gypsum thermal board with tapered
edges, fixing with adhesive to
smooth surfaces, joints taped
including one coat board
finish, 3mm thick

	Unit	Labour	Hours £	Mat'ls £	O & P £	Total £
40mm thick						
to walls						
over 300mm wide	m2	0.64	9.60	10.71	3.05	23.36
less than 300mm wide	m	0.23	3.45	3.71	1.07	8.23
to ceilings						
over 300mm wide	m2	0.68	10.20	10.71	3.14	24.05
less than 300mm wide	m	0.25	3.75	3.71	1.12	8.58
to beams						
over 300mm wide	m2	0.68	10.20	10.71	3.14	24.05
less than 300mm wide	m	0.25	3.75	3.71	1.12	8.58
to columns						
over 300mm wide	m2	0.68	10.20	10.71	3.14	24.05
less than 300mm wide	m	0.25	3.75	3.71	1.12	8.58
Raking cutting	m	0.28	4.20	-	0.63	4.83
Cutting around steel angles, joists, pipes and trunking						
over 2m girth	nr	0.32	4.80	-	0.72	5.52
less than 300mm girth	nr	0.22	3.30	-	0.50	3.80
300mm to 1m girth	nr	0.26	3.90	-	0.59	4.49
1m to 2m girth	nr	0.28	4.20	-	0.63	4.83

	Unit	Labour	Hours £	Mat'ls £	O & P £	Total £
Gypsum thermal board with tapered edges, fixing with adhesive to smooth surfaces, joints taped includingone coat board finish, 3mm thick						
to walls						
over 300mm wide						
less than 300mm wide	m2	0.68	10.20	11.21	3.21	24.62
to ceilings	m	0.24	3.60	3.79	1.11	8.50
over 300mm wide						
less than 300mm wide	m2	0.72	10.80	11.21	3.30	25.31
to beams	m	0.26	3.90	3.79	1.15	8.84
over 300mm wide						
less than 300mm wide	m2	0.72	10.80	11.21	3.30	25.31
to columns	m	0.26	3.90	3.79	1.15	8.84
over 300mm wide						
less than 300mm wide	m2	0.72	10.80	11.21	3.30	25.31
less than 300mm thick	m	0.26	3.90	3.79	1.15	8.84
Raking cutting	m	0.32	4.80	-	0.72	5.52
Cutting around steel angles, joists, pipes and trunking						
over 2m girth	nr	0.36	5.40	-	0.81	6.21
less than 300mm girth	nr	0.26	3.90	-	0.59	4.49
300mm to 1m girth	nr	0.28	4.20	-	0.63	4.83
1m to 2m girth	nr	0.32	4.80	-	0.72	5.52

	Unit	Labour	Hours £	Mat'ls £	O & P £	Total £
Supalux board with sanded finish, screwed to softwood with countersunk screws						
6mm thick						
to walls						
over 300mm wide	m2	0.40	6.00	14.14	3.02	23.16
less than 300mm wide	m	0.15	2.25	5.52	1.17	8.94
to ceilings						
over 300mm wide	m2	0.50	7.50	14.14	3.25	24.89
less than 300mm wide	m	0.18	2.70	5.52	1.23	9.45
to beams						
over 300mm wide	m2	0.60	9.00	14.14	3.47	26.61
less than 300mm wide	m	0.22	3.30	5.52	1.32	10.14
to columns						
over 300mm wide	m2	0.54	8.10	14.14	3.34	25.58
less than 300mm wide	m	0.18	2.70	5.52	1.23	9.45
Raking cutting	m	0.22	3.30	-	0.50	3.80
Cutting around steel angles, joists, pipes and trunking						
over 2m girth	nr	0.26	3.90	-	0.59	4.49
less than 300mm girth	nr	0.16	2.40	-	0.36	2.76
300mm to 1m girth	nr	0.18	2.70	-	0.41	3.11
1m to 2m girth	nr	0.22	3.30	-	0.50	3.80
Supalux board with sanded finish, screwed to softwood with countersunk screws						
9mm thick						
to walls						
over 300mm wide	m2	0.44	6.60	22.47	4.36	33.43
less than 300mm wide	m	0.17	2.55	8.40	1.64	12.59

	Unit	Labour	Hours £	Mat'ls £	O & P £	Total £
Supalux board 9mm thick (cont'd)						
to ceilings						
over 300mm wide	m2	0.54	8.10	22.47	4.59	35.16
less than 300mm wide	m	0.20	3.00	8.40	1.71	13.11
to beams						
over 300mm wide	m2	0.66	9.90	22.47	4.86	37.23
less than 300mm wide	m	0.24	3.60	8.40	1.80	13.80
to columns						
over 300mm wide	m2	0.58	8.70	22.47	4.68	35.85
less than 300mm wide	m	0.22	3.30	8.40	1.76	13.46
Raking cutting	m	0.24	3.60	-	0.54	4.14
Cutting around steel angles, joists, pipes and trunking						
over 2m girth	nr	0.28	4.20	-	0.63	4.83
less than 300mm girth	nr	0.18	2.70	-	0.41	3.11
300mm to 1m girth	nr	0.20	3.00	-	0.45	3.45
1m to 2m girth	nr	0.24	3.60	-	0.54	4.14
Supalux board with sanded finish, screwed to softwood with countersunk screws						
12mm thick						
to walls						
over 300mm wide	m2	0.48	7.20	37.21	6.66	51.07
less than 300mm wide	m	0.20	3.00	12.92	2.39	18.31
to ceilings						
over 300mm wide	m2	0.58	8.70	37.21	6.89	52.80
less than 300mm wide	m	0.22	3.30	12.92	2.43	18.65
to beams						
over 300mm wide	m2	0.70	10.50	37.21	7.16	54.87
less than 300mm wide	m	0.26	3.90	12.92	2.52	19.34

	Unit	Labour Hours	Mat'ls £	O & P £	Total £
to columns					
over 300mm wide	m2	0.62 9.30	37.21	6.98	53.49
less than 300mm wide	m	0.24 3.60	12.92	2.48	19.00
Raking cutting	m	0.26 3.90	-	0.59	4.49
Cutting around steel angles, joists, pipes and trunking					
over 2m girth	nr	0.30 4.50	-	0.68	5.18
less than 300mm girth	nr	0.20 3.00	-	0.45	3.45
300mm to 1m girth	nr	0.22 3.30	-	0.50	3.80
1m to 2m girth	nr	0.26 3.90	-	0.59	4.49

EXTERNAL RENDERING

Cement, lime and sand rendering (1:1:6) to concrete, brick or block walls

	Unit	Labour Hours	Mat'ls £	O & P £	Total £
12mm thick with wood floated finish					
over 300mm wide	m2	0.38 5.70	1.83	1.13	8.66
less than 300mm wide	m	0.18 2.70	0.70	0.51	3.91
12mm thick with wood dragged finish					
over 300mm wide	m2	0.44 6.60	1.83	1.26	9.69
less than 300mm wide	m	0.20 3.00	0.70	0.56	4.26
18mm thick with wood floated finish					
over 300mm wide	m2	0.54 8.10	2.72	1.62	12.44
less than 300mm wide	m	0.24 3.60	0.83	0.66	5.09

	Unit	Labour Hours	Mat'ls £	O & P £	Total £
18mm thick with wood dragged finish					
over 300mm wide	m2	0.60	9.00	2.72	1.76	13.48
less than 300mm wide	m	0.26	3.90	0.83	0.71	5.44

Cement, lime and sand waterproofed rendering (1:1:6) to concrete, brick or or block walls

	Unit	Labour Hours	Mat'ls £	O & P £	Total £
12mm thick with wood floated finish					
over 300mm wide	m2	0.38	5.70	2.11	1.17	8.98
less than 300mm wide	m	0.18	2.70	0.78	0.52	4.00
12mm thick with wood dragged finish					
over 300mm wide	m2	0.44	6.60	2.11	1.31	10.02
less than 300mm wide	m	0.20	3.00	0.78	0.57	4.35
18mm thick with wood floated finish					
over 300mm wide	m2	0.54	8.10	2.82	1.64	12.56
less than 300mm wide	m	0.24	3.60	0.97	0.69	5.26
18mm thick with wood dragged finish					
over 300mm wide	m2	0.60	9.00	2.82	1.77	13.59
less than 300mm wide	m	0.26	3.90	0.97	0.73	5.60

	Unit	Labour	Hours £	Mat'ls £	O & P £	Total £
Cement, lime and sand rendering (1:1:6) to concrete, brick or block walls with three coat Tyrolean finish						
12mm thick						
over 300mm wide	m2	0.68	10.20	4.08	2.14	16.42
less than 300mm wide	m	0.34	5.10	1.56	1.00	7.66
18mm thick						
over 300mm wide	m2	0.84	12.60	4.70	2.60	19.90
less than 300mm wide	m	0.36	5.40	2.01	1.11	8.52
Cement, lime and sand waterproofed rendering (1:1:6) to concrete, brick or or block walls with Tyrolean finish						
12mm thick						
over 300mm wide	m2	0.68	10.20	4.48	2.20	16.88
less than 300mm wide	m	0.34	5.10	1.62	1.01	7.73
18mm thick						
over 300mm wide	m2	0.84	12.60	5.20	2.67	20.47
less than 300mm wide	m	0.36	5.40	1.84	1.09	8.33
Cement, lime and sand rendering (1:1:6) to concrete, brick or block walls with Derbyshire Spar chippings						
12mm thick						
over 300mm wide	m2	1.10	16.50	2.08	2.79	21.37
less than 300mm wide	m	0.36	5.40	0.84	0.94	7.18

	Unit	Labour	Hours £	Mat'ls £	O & P £	Total £
18mm thick						
over 300mm wide	m2	1.14	17.10	2.41	2.93	22.44
less than 300mm wide	m	0.40	6.00	0.93	1.04	7.97

Cement, lime and sand
waterproofed rendering
(1:1:6) to concrete, brick or
or block walls with
Derbyshire Spar chippings

	Unit	Labour	Hours £	Mat'ls £	O & P £	Total £
12mm thick						
over 300mm wide	m2	1.10	16.50	2.43	2.84	21.77
less than 300mm wide	m	0.36	5.40	0.91	0.95	7.26
18mm thick						
over 300mm wide	m2	1.14	17.10	2.93	3.00	23.03
less than 300mm wide	m	0.40	6.00	1.19	1.08	8.27

REPAIR WORK

Break up defective sand
and cement screed and renew

	Unit	Labour	Hours £	Mat'ls £	O & P £	Total £
25mm thick	m2	0.40	6.00	3.15	1.37	10.52
32mm thick	m2	0.44	6.60	3.68	1.54	11.82
38mm thick	m2	0.48	7.20	3.90	1.67	12.77
50mm thick	m2	0.52	7.80	4.85	1.90	14.55
65mm thick	m2	0.56	8.40	6.15	2.18	16.73

Break up defective
granolithic screed and renew

	Unit	Labour	Hours £	Mat'ls £	O & P £	Total £
25mm thick	m2	0.50	7.50	4.63	1.82	13.95
32mm thick	m2	0.54	8.10	5.02	1.97	15.09
38mm thick	m2	0.58	8.70	5.41	2.12	16.23

	Unit	Labour Hours	Mat'ls £	O & P £	Total £
		£			
50mm thick	m2	0.62	9.30	6.16	2.32	17.78
65mm thick	m2	0.66	9.90	7.60	2.63	20.13

Hack off existing plaster
and replace with cement and
sand backing coat

| 13mm thick | m2 | 0.70 | 10.50 | 1.33 | 1.77 | 13.60 |
| 19mm thick | m2 | 0.78 | 11.70 | 2.38 | 2.11 | 16.19 |

Hack off existing plaster
and replace with lightweight
plaster, 11mm bonding and

| 2mm finish | m2 | 0.68 | 10.20 | 2.62 | 1.92 | 14.74 |

Hack off existing plaster
and replace with one coat
Universal' plaster

| 13mm thick | m2 | 0.50 | 7.50 | 3.21 | 1.61 | 12.32 |
| 19mm thick | m2 | 0.55 | 8.25 | 4.12 | 1.86 | 14.23 |

Hack off existing plaster
and replace with two coat
renovating plaster, 11mm

| undercoat and 2mm finish | m2 | 0.88 | 13.20 | 2.70 | 2.39 | 18.29 |

Hack up existing floor
covering and replace with
red quarry tiles

| 150 × 150 × 12.5mm thick | m2 | 1.10 | 16.50 | 31.82 | 7.25 | 55.57 |
| 200 × 200 × 19mm thick | m2 | 1.00 | 15.00 | 50.20 | 9.78 | 74.98 |

	Unit	Labour	Hours £	Mat'ls £	O & P £	Total £
Hack up existing floor covering and replace with brown quarry tiles						
150 × 150 × 12.5mm thick	m2	1.10	16.50	33.46	7.49	57.45
200 × 200 × 19mm thick	m2	1.00	15.00	52.63	10.14	77.77
Hack up existing floor covering and replace with vitrified ceramic tiles						
100 × 100 × 9mm thick	m2	1.20	18.00	26.36	6.65	51.01
150 × 150 × 12.5mm thick	m2	1.10	16.50	23.32	5.97	45.79
200 × 200 × 12.5mm thick	m2	1.00	15.00	21.12	5.42	41.54
Hack up existing floor covering and replace with terrazzo tile paving size						
300 × 300 × 28mm thick	m2	1.80	27.00	18.62	6.84	52.46
Hack up existing floor covering and replace with vinyl floor tiling						
300 × 300 × 2mm thick	m2	0.48	7.20	8.06	2.29	17.55
300 × 300 × 2.5mm thick	m2	0.58	8.70	8.48	2.58	19.76
Hack up existing floor covering and replace with cork floor tiling						
300 × 300 × 4mm thick	m2	0.56	8.40	24.65	4.96	38.01

	Unit	Labour	Hours £	Mat'ls £	O & P £	Total £
Hack up existing floor covering and replace with carpet tiling						
500 × 500	m2	0.58	8.70	11.65	3.05	23.40
Hack up existing floor covering and replace with vinyl sheeting						
2mm thick	m2	0.56	8.40	10.42	2.82	21.64
2.5mm thick	m2	0.60	9.00	14.27	3.49	26.76
Hack up existing floor covering and replace with linoleum						
2mm thick	m2	0.56	8.40	11.24	2.95	22.59
2.5mm thick	m2	0.60	9.00	12.54	3.23	24.77
Hack up existing floor covering and replace with woodblock flooring 25mm thick						
maple	m2	1.30	19.50	51.51	10.65	81.66
european oak	m2	1.30	19.50	43.52	9.45	72.47
iroko	m2	1.30	19.50	45.25	9.71	74.46
merbau	m2	1.30	19.50	45.25	9.71	74.46
sapele	m2	1.30	19.50	40.18	8.95	68.63

	Unit	Labour Hours	Mat'ls £	O & P £	Total £

Pull down existing lath and plaster ceiling and replace with gypsum plasterboard

	Unit	Labour Hours £	Mat'ls £	O & P £	Total £
9.5mm thick	m2	0.48	7.20	2.36	1.43	10.99
12.5mm thick	m2	0.05	0.75	2.68	0.51	3.94

Pull down existing lath and plaster ceiling and replace with gypsum plasterboard and one coat board finish 3mm thick

9.5mm thick	m2	0.74	11.10	3.70	2.22	17.02
12.5mm thick	m2	0.86	12.90	4.18	2.56	19.64

Pull down existing lath and plaster ceiling and replace with gypsum thermal board

30mm thick	m2	0.54	8.10	8.58	2.50	19.18
40mm thick	m2	0.58	8.70	9.02	2.66	20.38
50mm thick	m2	0.62	9.30	9.42	2.81	21.53

Pull down existing lath and plaster ceiling and replace with gypsum thermal board and one coat board finish 3mm thick

30mm thick	m2	0.86	12.90	10.14	3.46	26.50
40mm thick	m2	0.88	13.20	10.71	3.59	27.50
50mm thick	m2	0.92	13.80	11.24	3.76	28.80

Part Two

PROJECT COSTS – PLASTERING AND TILING

Project costs

PROJECT COSTS

This section gives the approximate costs of carrying out plastering
work in a typical house. The following assumptions have been made.

Living room:	1 door, 1 fireplace, 1 large window
Dining room:	2 doors, 1 large window
Bedroom	1 door, 1 average window
Kitchen	2 doors, fittings below dado level, 1 large window
WC	1 door, 1 small window
Bathroom	1 door, 1 average window, 1 cupboard

Ceiling heights: Ground floor 2.4 metres and First floor 2.2 metres.

The following rounded-off quantities for plastering ceilings and walls.

	Ceilings	Walls
Living room		
3.6 x 3.0m	11m2	24m2
3.6 x 3.6m	13m2	26m2
4.2 x 3.0m	13m2	26m2
4.2 x 3.6m	15m2	29m2
4.2 x 4.2m	18m2	32m2
4.8 x 3.0m	15m2	29m2
4.8 x 3.6m	17m2	32m2
4.8 x 4.2m	20m2	34m2
4.8 x 4.8m	23m2	37m2
Dining room		
3.0 x 3.0m	9m2	20m2
3.6 x 3.0m	11m2	24m2
3.6 x 3.6m	13m2	26m2
4.2 x 3.0m	13m2	26m2
4.2 x 3.6m	15m2	29m2
4.2 x 4.2m	18m2	32m2
4.8 x 3.0m	15m2	29m2
4.8 x 3.6m	17m2	32m2
4.8 x 4.2m	20m2	24m2

	Ceilings	Walls
Bedroom		
2.4 x 2.4m	6m2	17m2
2.4 x 3.0m	7m2	20m2
2.4 x 3.6m	9m2	22m2
3.0 x 3.0m	9m2	22m2
3.0 x 3.6m	11m2	25m2
3.6 x 3.6m	13m2	27m2
3.6 x 4.2m	15m2	30m2
3.6 x 4.8m	17m2	32m2
4.2 x 4.2m	18m2	32m2
4.2 x 4.8m	20m2	35m2
Kitchen		
1.8 x 3.0m	6m2	7m2
1.8 x 3.6m	7m2	10m2
2.4 x 3.0m	7m2	10m2
2.4 x 3.6m	9m2	11m2
3.0 x 3.0m	9m2	11m2
3.0 x 3.6m	11m2	11m2
3.0 x 4.2m	13m2	11m2
3.6 x 3.6m	13m2	13m2
3.6 x 4.2m	15m2	15m2
WC		
1.0 x 1.5m	2m2	8m2
1.2 x 1.5m	2m2	9m2
1.2 x 1.8m	2m2	10m2
1.3 x 1.5m	2m2	9m2
1.3 x 1.8m	2m2	10m2
Bathroom		
1.8 x 2.4m	4m2	23m2
1.8 x 3.0m	5m2	25m2
2.1 x 3.4m	5m2	23m2

Bathroom (cont'd)

2.4 x 2.4m	6m2	25m2
2.4 x 3.0m	7m2	28m2

The specification for the work carried out is:

Ceilings: Plasterboard 12.5mm thick and one coat board finish

Walls: One coat 'Universal' plaster 19mm thick

The cost of the work in each room is assessed by multiplying the cost data in Unit Rates by the quantities listed above. Note that all the figures are rounded off.

Living room   £

3.6 x 3.0m	425
3.6 x 3.6m	460
4.2 x 3.0m	460
4.2 x 3.6m	540
4.2 x 4.2m	620
4.8 x 3.0m	540
4.8 x 3.6m	600
4.8 x 4.2m	670
4.8 x 4.8m	750

Dining room

3.0 x 3.0m	525
3.6 x 3.0m	425
3.6 x 3.6m	460
4.2 x 3.0m	460
4.2 x 3.6m	540
4.2 x 4.2m	620
4.8 x 3.0m	540
4.8 x 3.6m	600
4.8 x 4.2m	670

Bedroom

2.4 x 2.4m	275
2.4 x 3.0m	320

Bedroom (cont'd)

2.4 x 3.6m	375
3.0 x 3.0m	375
3.0 x 3.6m	435
3.6 x 3.6m	470
3.6 x 4.2m	550
3.6 x 4.8m	600
4.2 x 4.2m	620
4.2 x 4.8m	680

Kitchen

1.8 x 3.0m	170
1.8 x 3.6m	215
2.4 x 3.0m	215
2.4 x 3.6m	255
3.0 x 3.0m	255
3.0 x 3.6m	290
3.0 x 4.2m	305
3.6 x 3.6m	325
3.6 x 4.2m	395

WC

1.0 x 1.5m	115
1.2 x 1.5m	125
1.2 x 1.8m	135
1.3 x 1.5m	125
1.3 x 1.8m	135

Bathroom

1.8 x 2.4m	305
1.8 x 3.0m	340
2.1 x 2.4m	320
2.4 x 2.4m	360
2.4 x 3.0m	405
2.4 x 3.6m	455

Part Three

UNIT RATES – PAINTING AND WALLPAPERING

Internal work

        Preparatory work

        Primers

        Undercoats

        Finishing coats

        Wallpapering

External work

        Preparatory work

        Primers

        Undercoats

        Finishing coats

	Unit	Labour Hours £	Mat'ls £	O & P £	Total £

INTERNAL WORK

The unit rates in this section
include all the necessary
preparatory work involved
in treating existing surfaces
to receive new paintwork.
This includes cutting out and
filling cracks, knotting and
stopping where necessary
and a minimum allowance
for patch priming.

The rates are based upon
brush application and working
in normal conditions on
surfaces that could be described
as in 'average condition'.

The following adjustments
apply where necessary:

working at heights over 3m
   add 15% to labour

working on surfaces in poor
condition
   add 10% to labour and
   materials

working in cramped conditions
   add 15% to labour

spraying
   deduct 30% of labour and
   add 15% to materials

	Unit	Labour Hours	Mat'ls £	O & P £	Total £

PREPARATORY WORK

Wash down painted surfaces
over 300mm girth, stop
cracks and rub down

	Unit	Labour Hours	Mat'ls £	O & P £	Total £
brickwork					
walls	m2	0.14	2.10	0.06	0.32	2.48
walls in staircase areas	m2	0.16	2.40	0.06	0.37	2.83
blockwork					
walls	m2	0.15	2.25	0.06	0.35	2.66
walls in staircase areas	m2	0.17	2.55	0.06	0.39	3.00
concrete					
walls	m2	0.14	2.10	0.06	0.32	2.48
walls in staircase areas	m2	0.16	2.40	0.06	0.37	2.83
ceilings	m2	0.17	2.55	0.06	0.39	3.00
ceilings in staircase areas	m2	0.19	2.85	0.06	0.44	3.35
plastered					
walls	m2	0.12	1.80	0.06	0.28	2.14
walls in staircase areas	m2	0.14	2.10	0.06	0.32	2.48
ceilings	m2	0.15	2.25	0.06	0.35	2.66
ceilings in staircase areas	m2	0.17	2.55	0.06	0.39	3.00
embossed paper					
walls	m2	0.13	1.95	0.06	0.30	2.31
walls in staircase areas	m2	0.14	2.10	0.06	0.32	2.48
ceilings	m2	0.17	2.55	0.06	0.39	3.00
ceilings in staircase areas	m2	0.19	2.85	0.06	0.44	3.35

	Unit	Labour	Hours £	Mat'ls £	O & P £	Total £

Wash down emulsioned surfaces
over 300mm girth, stop
cracks and rub down

	Unit	Labour	Hours £	Mat'ls £	O & P £	Total £
brickwork						
walls	m2	0.12	1.80	0.06	0.28	2.14
walls in staircase areas	m2	0.14	2.10	0.06	0.32	2.48
blockwork						
walls	m2	0.13	1.95	0.06	0.30	2.31
walls in staircase areas	m2	0.15	2.25	0.06	0.35	2.66
concrete						
walls	m2	0.12	1.80	0.06	0.28	2.14
walls in staircase areas	m2	0.14	2.10	0.06	0.32	2.48
ceilings	m2	0.15	2.25	0.06	0.35	2.66
ceilings in staircase areas	m2	0.17	2.55	0.06	0.39	3.00
plastered						
walls	m2	0.12	1.80	0.06	0.28	2.14
walls in staircase areas	m2	0.14	2.10	0.06	0.32	2.48
ceilings	m2	0.13	1.95	0.06	0.30	2.31
ceilings in staircase areas	m2	0.15	2.25	0.06	0.35	2.66
embossed paper						
walls	m2	0.11	1.65	0.06	0.26	1.97
walls in staircase areas	m2	0.12	1.80	0.06	0.28	2.14
ceilings	m2	0.15	2.25	0.06	0.35	2.66
ceilings in staircase areas	m2	0.17	2.55	0.06	0.39	3.00

	Unit	Labour Hours	Mat'ls £	O & P £	Total £

Wash down oil-painted surfaces
over 300mm girth, stop
cracks and rub down

	Unit	Labour Hours	Mat'ls £	O & P £	Total £
brickwork					
walls	m2	0.10	1.50	0.06	0.23	1.79
walls in staircase areas	m2	0.12	1.80	0.06	0.28	2.14
blockwork					
walls	m2	0.11	1.65	0.06	0.26	1.97
walls in staircase areas	m2	0.13	1.95	0.06	0.30	2.31
concrete					
walls	m2	0.10	1.50	0.06	0.23	1.79
walls in staircase areas	m2	0.12	1.80	0.06	0.28	2.14
ceilings	m2	0.13	1.95	0.06	0.30	2.31
ceilings in staircase areas	m2	0.15	2.25	0.06	0.35	2.66
plastered					
walls	m2	0.10	1.50	0.06	0.23	1.79
walls in staircase areas	m2	0.12	1.80	0.06	0.28	2.14
ceilings	m2	0.11	1.65	0.06	0.26	1.97
ceilings in staircase areas	m2	0.13	1.95	0.06	0.30	2.31
embossed paper					
walls	m2	0.09	1.35	0.06	0.21	1.62
walls in staircase areas	m2	0.10	1.50	0.06	0.23	1.79
ceilings	m2	0.13	1.95	0.06	0.30	2.31
ceilings in staircase areas	m2	0.15	2.25	0.06	0.35	2.66

	Unit	Labour	Hours £	Mat'ls £	O & P £	Total £
Wash down oil-painted wood surfaces and rub down						
general surfaces						
over 300mm girth	m2	0.28	4.20	-	0.63	4.83
not exceeding 300mm girth	m	0.10	1.50	-	0.23	1.73
150 to 300mm girth	m	0.08	1.20	-	0.18	1.38
isolated areas not exceeding 0.5m2	nr	0.14	2.10	-	0.32	2.42
windows, screens glazed doors and the like						
panes area not exceeding 0.2m2	m2	0.40	6.00	-	0.90	6.90
panes area 0.1 to 0.5m2	m2	0.37	5.55	-	0.83	6.38
panes area 0.5 to 1m2	m2	0.34	5.10	-	0.77	5.87
panes area exceeding 1m2	m2	0.30	4.50	-	0.68	5.18
Wash down oil-painted metal surfaces and rub down						
general surfaces						
over 300mm girth	m2	0.28	4.20	-	0.63	4.83
not exceeding 300mm girth	m	0.10	1.50	-	0.23	1.73
150 to 300mm girth	m	0.08	1.20	-	0.18	1.38
isolated areas not exceeding 0.5m2	nr	0.14	2.10	-	0.32	2.42

	Unit	Labour Hours	£	Mat'ls £	O & P £	Total £
Wash down oil-painted metal surfaces (cont'd)						
windows, screens glazed doors and the like						
panes area not exceeding 0.2m2	m2	0.40	6.00	-	0.90	6.90
panes area 0.1 to 0.5m2	m2	0.37	5.55	-	0.83	6.38
panes area 0.5 to 1m2	m2	0.34	5.10	-	0.77	5.87
panes area exceeding 1m2	m2	0.30	4.50	-	0.68	5.18
structural metalwork, general surfaces						
over 300mm girth	m2	0.35	5.25	-	0.79	6.04
not exceeding 150mm girth	m	0.06	0.90	-	0.14	1.04
150 to 300mm girth	m	0.12	1.80	-	0.27	2.07
isolated areas not exceeding 0.5m2	nr	0.18	2.70	-	0.41	3.11
structural metalwork, roof truss members						
over 300mm girth	m2	0.50	7.50	-	1.13	8.63
not exceeding 150mm girth	m	0.08	1.20	-	0.18	1.38
150 to 300mm girth	m	0.16	2.40	-	0.36	2.76
isolated areas not exceeding 0.5m2	nr	0.25	3.75	-	0.56	4.31
radiators, panel type						
over 300mm girth	m2	0.30	4.50	-	0.68	5.18
not exceeding 300mm girth	m	0.12	1.80	-	0.27	2.07
150 to 300mm girth	m	0.10	1.50	-	0.23	1.73
isolated areas not exceeding 0.5m2	nr	0.15	2.25	-	0.34	2.59

	Unit	Labour	Hours £	Mat'ls £	O & P £	Total £
radiators, column type						
over 300mm girth	m2	0.40	6.00	-	0.90	6.90
not exceeding 300mm girth	m	0.14	2.10	-	0.32	2.42
150 to 300mm girth	m	0.12	1.80	-	0.27	2.07
isolated areas not exceeding 0.5m2	nr	0.20	3.00	-	0.45	3.45

Wash down oil-painted wood surfaces, remove paint with chemical stripper and rub down

	Unit	Labour	Hours £	Mat'ls £	O & P £	Total £
general surfaces						
over 300mm girth	m2	0.80	12.00	2.60	2.19	16.79
not exceeding 300mm girth	m	0.30	4.50	0.98	0.82	6.30
150 to 300mm girth	m	0.26	3.90	0.65	0.68	5.23
isolated areas not exceeding 0.5m2	nr	0.40	6.00	1.31	1.10	8.41
windows, screens glazed doors and the like						
panes area not exceeding 0.2m2	m2	1.00	15.00	2.17	2.58	19.75
panes area 0.1 to 0.5m2	m2	0.95	14.25	2.06	2.45	18.76
panes area 0.5 to 1m2	m2	0.90	13.50	1.94	2.32	17.76
panes area exceeding 1m2	m2	0.85	12.75	1.86	2.19	16.80

	Unit	Labour Hours	Mat'ls £	O & P £	Total £
Wash down oil-painted metal surfaces, remove paint with chemical stripper and rub down					
general surfaces					
over 300mm girth	m2	0.80	12.00	2.60	2.19	16.79
not exceeding 300mm girth	m	0.30	4.50	0.98	0.82	6.30
150 to 300mm girth	m	0.26	3.90	0.65	0.68	5.23
isolated areas not exceeding 0.5m2	nr	0.40	6.00	1.31	1.10	8.41
windows, screens glazed doors and the like					
panes area not exceeding 0.2m2	m2	1.00	15.00	2.17	2.58	19.75
panes area 0.1 to 0.5m2	m2	0.95	14.25	2.06	2.45	18.76
panes area 0.5 to 1m2	m2	0.80	12.00	1.94	2.09	16.03
panes area exceeding 1m2	m2	0.85	12.75	1.86	2.19	16.80
structural metalwork, general surfaces					
over 300mm girth	m2	0.80	12.00	2.60	2.19	16.79
not exceeding 150mm girth	m	0.15	2.25	0.98	0.48	3.71
150 to 300mm girth	m	0.40	6.00	0.65	1.00	7.65
isolated areas not exceeding 0.5m2	nr	0.40	6.00	1.31	1.10	8.41

	Unit	Labour	Hours £	Mat'ls £	O & P £	Total £
structural metalwork, roof truss members						
over 300mm girth	m2	1.80	27.00	2.60	4.44	34.04
not exceeding 300mm girth	m	0.65	9.75	0.98	1.61	12.34
150 to 300mm girth	m	0.58	8.70	0.65	1.40	10.75
isolated areas not exceeding 0.5m2	nr	0.90	13.50	1.31	2.22	17.03
radiators, panel type						
over 300mm girth	m2	1.00	15.00	2.60	2.64	20.24
not exceeding 150mm girth	m	0.14	2.10	0.98	0.46	3.54
150 to 300mm girth	m	0.30	4.50	0.65	0.77	5.92
isolated areas not exceeding 0.5m2	nr	0.50	7.50	1.31	1.32	10.13
radiators, column type						
over 300mm girth	m2	1.20	18.00	2.60	3.09	23.69
not exceeding 150mm girth	m	0.18	2.70	0.98	0.55	4.23
150 to 300mm girth	m	0.38	5.70	0.65	0.95	7.30
isolated areas not exceeding 0.5m2	nr	0.60	9.00	1.31	1.55	11.86

Burn off paint from wood surfaces and rub down

	Unit	Labour	Hours £	Mat'ls £	O & P £	Total £
general surfaces						
over 300mm girth	m2	1.05	15.75	-	2.36	18.11
not exceeding 300mm girth	m	0.30	4.50	-	0.68	5.18
150 to 300mm girth	m	0.40	6.00	-	0.90	6.90
isolated areas not exceeding 0.5m2	nr	0.50	7.50	-	1.13	8.63

	Unit	Labour Hours	Mat'ls £	O & P £	Total £
Burn off paint from wood surfaces and rub down (cont'd)					
windows, screens glazed doors and the like					
panes area not exceeding 0.2m2	m2	1.70	25.50	- 3.83	29.33
panes area 0.1 to 0.5m2	m2	1.60	24.00	- 3.60	27.60
panes area 0.5 to 1m2	m2	1.50	22.50	- 3.38	25.88
panes area exceeding 1m2	m2	1.40	21.00	- 3.15	24.15
Burn off paint from metal surfaces and rub down					
general surfaces					
over 300mm girth	m2	1.05	15.75	- 2.36	18.11
not exceeding 300mm girth	m	0.30	4.50	- 0.68	5.18
150 to 300mm girth	m	0.40	6.00	- 0.90	6.90
isolated areas not exceeding 0.5m2	nr	0.50	7.50	- 1.13	8.63
windows, screens glazed doors and the like					
panes area not exceeding 0.2m2	m2	1.70	25.50	- 3.83	29.33
panes area 0.1 to 0.5m2	m2	1.60	24.00	- 3.60	27.60
panes area 0.5 to 1m2	m2	1.50	22.50	- 3.38	25.88
panes area exceeding 1m2	m2	1.40	21.00	- 3.15	24.15

	Unit	Labour	Hours £	Mat'ls £	O & P £	Total £
structural metalwork, general surfaces						
over 300mm girth	m2	1.05	15.75	-	2.36	18.11
not exceeding 150mm girth	m	0.16	2.40	-	0.36	2.76
150 to 300mm girth	m	0.40	6.00	-	0.90	6.90
isolated areas not exceeding 0.5m2	nr	0.50	7.50	-	1.13	8.63
structural metalwork, roof truss members						
over 300mm girth	m2	1.80	27.00	-	4.05	31.05
not exceeding 150mm girth	m	0.30	4.50	-	0.68	5.18
150 to 300mm girth	m	0.58	8.70	-	1.31	10.01
isolated areas not exceeding 0.5m2	nr	0.90	13.50	-	2.03	15.53
radiators, panel type						
over 300mm girth	m2	1.05	15.75	-	2.36	18.11
not exceeding 150mm girth	m	0.15	2.25	-	0.34	2.59
150 to 300mm girth	m	0.40	6.00	-	0.90	6.90
isolated areas not exceeding 0.5m2	nr	0.50	7.50	-	1.13	8.63
radiators, column type						
over 300mm girth	m2	1.20	18.00	-	2.70	20.70
not exceeding 150mm girth	m	0.18	2.70	-	0.41	3.11
150 to 300mm girth	m	0.38	5.70	-	0.86	6.56
isolated areas not exceeding 0.5m2	nr	0.60	9.00	-	1.35	10.35

	Unit	Labour Hours	£	Mat'ls £	O & P £	Total £

PRIMERS

One coat emulsion paint as
primer on surfaces over
300mm girth

	Unit	Labour Hours	Mat'ls £	O & P £	Total £
brickwork walls	m2	0.10	1.50	0.62	0.32	2.44
brickwork walls in staircase areas	m2	0.12	1.80	0.62	0.36	2.78
blockwork walls	m2	0.12	1.80	0.81	0.39	3.00
blockwork walls in staircase areas	m2	0.14	2.10	0.81	0.44	3.35
concrete walls	m2	0.09	1.35	0.49	0.28	2.12
concrete walls in staircase areas	m2	0.11	1.65	0.49	0.32	2.46
concrete ceilings	m2	0.13	1.95	0.49	0.37	2.81
concrete ceilings in staircase areas	m2	0.15	2.25	0.49	0.41	3.15
plastered walls	m2	0.08	1.20	0.49	0.25	1.94
plastered walls in staircase areas	m2	0.10	1.50	0.49	0.30	2.29
plastered ceilings	m2	0.10	1.50	0.49	0.30	2.29
plastered ceilings in staircase areas	m2	0.12	1.80	0.49	0.34	2.63
embossed paper	m2	0.08	1.20	0.49	0.25	1.94
embossed paper in staircase areas	m2	0.10	1.50	0.49	0.30	2.29

One coat alkali-resisting
primer on surfaces over
300mm girth

	Unit	Labour Hours	Mat'ls £	O & P £	Total £
brickwork walls	m2	0.10	1.50	1.53	0.45	3.48
brickwork walls in staircase areas	m2	0.12	1.80	1.53	0.50	3.83
blockwork walls	m2	0.12	1.80	2.15	0.59	4.54

	Unit	Labour Hours £	Mat'ls £	O & P £	Total £
blockwork walls in staircase areas	m2	0.14 2.10	2.15	0.64	4.89
concrete walls	m2	0.09 1.35	1.18	0.38	2.91
concrete walls in staircase areas	m2	0.11 1.65	1.18	0.42	3.25
concrete ceilings	m2	0.13 1.95	1.18	0.47	3.60
concrete ceilings in staircase areas	m2	0.15 2.25	1.18	0.51	3.94
plastered walls	m2	0.08 1.20	1.18	0.36	2.74
plastered walls in staircase areas	m2	0.10 1.50	1.18	0.40	3.08
plastered ceilings	m2	0.10 1.50	1.18	0.40	3.08
plastered ceilings in staircase areas	m2	0.12 1.80	1.18	0.45	3.43
embossed paper	m2	0.08 1.20	1.18	0.36	2.74
embossed paper in staircase areas	m2	0.10 1.50	1.18	0.40	3.08

One coat wood primer on wood

general surfaces					
over 300mm girth	m2	0.18 2.70	0.79	0.52	4.01
not exceeding 150mm girth	m	0.06 0.90	0.14	0.16	1.20
150 to 300mm girth	m	0.10 1.50	0.32	0.27	2.09
isolated areas not exceeding 0.5m2	nr	0.12 1.80	0.40	0.33	2.53

windows, screens glazed doors and the like					
panes area not exceeding 0.2m2	m2	0.35 5.25	0.42	0.85	6.52
panes area 0.1 to 0.5m2	m2	0.30 4.50	0.38	0.73	5.61
panes area 0.5 to 1m2	m2	0.25 3.75	0.33	0.61	4.69
panes area exceeding 1m2	m2	0.22 3.30	0.29	0.54	4.13

	Unit	Labour Hours £	Mat'ls £	O & P £	Total £
One coat wood primer on wood (cont'd)					
frames and linings					
over 300mm girth	m2	0.26	3.90	0.79	0.70	5.39
not exceeding 150mm girth	m	0.06	0.90	0.14	0.16	1.20
150 to 300mm girth	m	0.10	1.50	0.32	0.27	2.09
skirtings and rails					
over 300mm girth	m2	0.28	4.20	0.79	0.75	5.74
not exceeding 150mm girth	m	0.12	1.80	0.14	0.29	2.23
150 to 300mm girth	m	0.16	2.40	0.32	0.41	3.13
One coat aluminium primer on wood					
general surfaces					
over 300mm girth	m2	0.18	2.70	1.44	0.62	4.76
not exceeding 150mm girth	m	0.06	0.90	0.26	0.17	1.33
150 to 300mm girth	m	0.10	1.50	0.49	0.30	2.29
isolated areas not exceeding 0.5m2	nr	0.12	1.80	0.72	0.38	2.90
windows, screens glazed doors and the like					
panes area not exceeding 0.2m2	m2	0.35	5.25	0.72	0.90	6.87
panes area 0.1 to 0.5m2	m2	0.30	4.50	0.67	0.78	5.95
panes area 0.5 to 1m2	m2	0.25	3.75	0.63	0.66	5.04
panes area exceeding 1m2	m2	0.22	3.30	0.58	0.58	4.46

	Unit	Labour	Hours £	Mat'ls £	O & P £	Total £
frames and linings						
over 300mm girth	m2	0.26	3.90	1.44	0.80	6.14
not exceeding 150mm girth	m	0.06	0.90	0.26	0.17	1.33
150 to 300mm girth	m	0.10	1.50	0.47	0.30	2.27
skirtings and rails						
over 300mm girth	m2	0.28	4.20	1.44	0.85	6.49
not exceeding 150mm girth	m	0.12	1.80	0.26	0.31	2.37
150 to 300mm girth	m	0.16	2.40	0.47	0.43	3.30

One coat acrylic primer on wood

	Unit	Labour	Hours £	Mat'ls £	O & P £	Total £
general surfaces						
over 300mm girth	m2	0.18	2.70	0.79	0.52	4.01
not exceeding 150mm girth	m	0.06	0.90	0.14	0.16	1.20
150 to 300mm girth	m	0.10	1.50	0.32	0.27	2.09
isolated areas not exceeding 0.5m2	nr	0.12	1.80	0.40	0.33	2.53
windows, screens glazed doors and the like						
panes area not exceeding 0.2m2	m2	0.35	5.25	0.42	0.85	6.52
panes area 0.1 to 0.5m2	m2	0.30	4.50	0.38	0.73	5.61
panes area 0.5 to 1m2	m2	0.25	3.75	0.33	0.61	4.69
panes area exceeding 1m2	m2	0.22	3.30	0.29	0.54	4.13
frames and linings						
over 300mm girth	m2	0.26	3.90	0.79	0.70	5.39
not exceeding 150mm girth	m	0.06	0.90	0.14	0.16	1.20
150 to 300mm girth	m	0.10	1.50	0.32	0.27	2.09

	Unit	Labour Hours	Mat'ls £	O & P £	Total £
		£			

One coat acrylic primer (cont'd)

skirtings and rails					
over 300mm girth	m2	0.28	4.20	0.79	0.12	5.11
not exceeding 150mm girth	m	0.12	1.80	0.14	0.02	1.96
150 to 300mm girth	m	0.16	2.40	0.32	0.32	3.04

One coat zinc chromate primer on metal

general surfaces					
over 300mm girth	m2	0.18	2.70	0.61	0.09	3.40
not exceeding 150mm girth	m	0.06	0.90	0.11	0.02	1.03
150 to 300mm girth	m	0.10	1.50	0.21	0.17	1.88
isolated areas not exceeding 0.5m2	nr	0.12	1.80	0.31	0.05	2.16

windows, screens glazed doors and the like					
panes area not exceeding 0.2m2	m2	0.35	5.25	0.31	0.05	5.61
panes area 0.1 to 0.5m2	m2	0.30	4.50	0.26	0.83	6.34
panes area 0.5 to 1m2	m2	0.25	3.75	0.22	0.71	4.68
panes area exceeding 1m2	m2	0.22	3.30	0.18	0.59	4.07

structural metalwork, general surfaces					
over 300mm girth	m2	0.30	4.50	0.63	0.09	5.22
not exceeding 150mm girth	m	0.05	0.75	0.11	0.02	0.88
150 to 300mm girth	m	0.10	1.50	0.21	0.14	1.85
isolated areas not exceeding 0.5m2	nr	0.15	2.25	0.31	0.05	2.61

	Unit	Labour	Hours £	Mat'ls £	O & P £	Total £
structural metalwork, roof truss members						
over 300mm girth	m2	0.40	6.00	0.63	0.09	0.72
not exceeding 150mm girth	m	0.08	1.20	0.11	0.20	1.51
150 to 300mm girth	m	0.06	0.90	0.21	0.17	1.28
isolated areas not exceeding 0.5m2	nr	0.20	3.00	0.31	0.50	3.81
radiators, panel type						
over 300mm girth	m2	0.18	2.70	0.63	0.50	3.83
not exceeding 150mm girth	m	0.06	0.90	0.11	0.15	1.16
150 to 300mm girth	m	0.10	1.50	0.21	0.26	1.97
isolated areas not exceeding 0.5m2	nr	0.12	1.80	0.31	0.32	2.43
radiators, column type						
over 300mm girth	m2	0.30	4.50	0.63	0.77	5.90
not exceeding 150mm girth	m	0.05	0.75	0.11	0.13	0.99
150 to 300mm girth	m	0.10	1.50	0.21	0.26	1.97
isolated areas not exceeding 0.5m2	nr	0.15	2.25	0.31	0.38	2.94
One coat metal red oxide primer on metal						
general surfaces						
over 300mm girth	m2	0.18	2.70	0.56	0.49	3.75
not exceeding 150mm girth	m	0.06	0.90	0.10	0.15	1.15
150 to 300mm girth	m	0.10	1.50	0.19	0.25	1.94
isolated areas not exceeding 0.5m2	nr	0.12	1.80	0.29	0.31	2.40

	Unit	Labour Hours	Mat'ls £	O & P £	Total £

One coat metal red oxide primer (cont'd)

windows, screens glazed doors and the like

	Unit	Labour Hours	Mat'ls £	O & P £	Total £
panes area not exceeding 0.2m2	m2	0.35	5.25	0.28	0.83	6.36
panes area 0.1 to 0.5m2	m2	0.30	4.50	0.24	0.71	5.45
panes area 0.5 to 1m2	m2	0.25	3.75	0.20	0.59	4.54
panes area exceeding 1m2	m2	0.22	3.30	0.16	0.52	3.98

structural metalwork, general surfaces

	Unit	Labour Hours	Mat'ls £	O & P £	Total £
over 300mm girth	m2	0.30	4.50	0.56	0.76	5.82
not exceeding 150mm girth	m	0.05	0.75	0.10	0.13	0.98
150 to 300mm girth	m	0.10	1.50	0.19	0.25	1.94
isolated areas not exceeding 0.5m2	nr	0.15	2.25	0.29	0.38	2.92

structural metalwork, roof truss members

	Unit	Labour Hours	Mat'ls £	O & P £	Total £
over 300mm girth	m2	0.40	6.00	0.56	0.98	7.54
not exceeding 150mm girth	m	0.08	1.20	0.10	0.20	1.50
150 to 300mm girth	m	0.06	0.90	0.19	0.16	1.25
isolated areas not exceeding 0.5m2	nr	0.20	3.00	0.29	0.49	3.78

radiators, panel type

	Unit	Labour Hours	Mat'ls £	O & P £	Total £
over 300mm girth	m2	0.18	2.70	0.56	0.49	3.75
not exceeding 150mm girth	m	0.06	0.90	0.10	0.15	1.15
150 to 300mm girth	m	0.10	1.50	0.19	0.25	1.94
isolated areas not exceeding 0.5m2	nr	0.12	1.80	0.29	0.31	2.40

	Unit	Labour	Hours £	Mat'ls £	O & P £	Total £
radiators, column type						
over 300mm girth	m2	0.30	4.50	0.56	0.76	5.82
not exceeding 150mm girth	m	0.05	0.75	0.10	0.13	0.98
150 to 300mm girth	m	0.10	1.50	0.19	0.25	1.94
isolated areas not exceeding 0.5m2	nr	0.15	2.25	0.29	0.38	2.92

One coat rubber paint as primer on surfaces over 300mm girth

	Unit	Labour	Hours £	Mat'ls £	O & P £	Total £
brickwork walls	m2	0.20	3.00	2.25	0.79	6.04
brickwork walls in staircase areas	m2	0.22	3.30	2.25	0.83	6.38
blockwork walls	m2	0.22	3.30	2.81	0.92	7.03
blockwork walls in staircase areas	m2	0.24	3.60	2.81	0.96	7.37
concrete walls	m2	0.18	2.70	1.84	0.68	5.22
concrete walls in staircase areas	m2	0.20	3.00	1.84	0.73	5.57
concrete ceilings	m2	0.22	3.30	1.84	0.77	5.91
concrete ceilings in staircase areas	m2	0.24	3.60	1.84	0.82	6.26
plastered walls	m2	0.18	2.70	1.84	0.68	5.22
plastered walls in staircase areas	m2	0.20	3.00	1.84	0.73	5.57
plastered ceilings	m2	0.22	3.30	1.84	0.77	5.91
plastered ceilings in staircase areas	m2	0.24	3.60	1.84	0.82	6.26

	Unit	Labour	Hours £	Mat'ls £	O & P £	Total £

UNDERCOATS

One coat emulsion paint on primed surfaces over 300mm girth

	Unit	Labour	Hours	Mat'ls	O & P	Total
brickwork walls	m2	0.08	1.20	0.57	0.27	2.04
brickwork walls in staircase areas	m2	0.10	1.50	0.57	0.31	2.38
blockwork walls	m2	0.10	1.50	0.77	0.34	2.61
blockwork walls in staircase areas	m2	0.12	1.80	0.77	0.39	2.96
concrete walls	m2	0.07	1.05	0.44	0.22	1.71
concrete walls in staircase areas	m2	0.09	1.35	0.44	0.27	2.06
concrete ceilings	m2	0.11	1.65	0.44	0.31	2.40
concrete ceilings in staircase areas	m2	0.03	0.45	0.44	0.13	1.02
plastered walls	m2	0.06	0.90	0.44	0.20	1.54
plastered walls in staircase areas	m2	0.08	1.20	0.44	0.25	1.89
plastered ceilings	m2	0.08	1.20	0.44	0.25	1.89
plastered ceilings in staircase areas	m2	0.10	1.50	0.44	0.29	2.23
embossed paper	m2	0.06	0.90	0.44	0.20	1.54
embossed paper in staircase areas	m2	0.08	1.20	0.44	0.25	1.89

One coat white oil-based undercoat on primed surfaces over 300mm girth

	Unit	Labour	Hours	Mat'ls	O & P	Total
brickwork walls	m2	0.08	1.20	0.44	0.25	1.89
brickwork walls in staircase areas	m2	0.10	1.50	0.44	0.29	2.23
blockwork walls	m2	0.10	1.50	0.44	0.29	2.23

	Unit	Labour Hours	Mat'ls £	O & P £	Total £
blockwork walls in staircase areas	m2	0.12	1.80	0.44	0.34	2.58
concrete walls	m2	0.11	1.65	0.44	0.31	2.40
concrete walls in staircase areas	m2	0.09	1.35	0.44	0.27	2.06
concrete ceilings	m2	0.11	1.65	0.44	0.31	2.40
concrete ceilings in staircase areas	m2	0.13	1.95	0.44	0.36	2.75
plastered walls	m2	0.06	0.90	0.44	0.20	1.54
plastered walls in staircase areas	m2	0.08	1.20	0.44	0.25	1.89
plastered ceilings	m2	0.08	1.20	0.44	0.25	1.89
plastered ceilings in staircase areas	m2	0.10	1.50	0.44	0.29	2.23
embossed paper	m2	0.06	0.90	0.44	0.20	1.54
embossed paper in staircase areas	m2	0.08	1.20	0.44	0.25	1.89
general surfaces						
over 300mm girth	m2	0.16	2.40	0.44	0.43	3.27
not exceeding 150mm girth	m	0.05	0.75	0.08	0.12	0.95
150 to 300mm girth	m	0.09	1.35	0.16	0.23	1.74
isolated areas not exceeding 0.5m2	nr	0.11	1.65	0.24	0.28	2.17
windows, screens glazed doors and the like						
panes area not exceeding 0.2m2	m2	0.33	4.95	0.30	0.79	6.04
panes area 0.1 to 0.5m2	m2	0.28	4.20	0.26	0.67	5.13
panes area 0.5 to 1m2	m2	0.23	3.45	0.22	0.55	4.22
panes area exceeding 1m2	m2	0.20	3.00	0.18	0.48	3.66

	Unit	Labour	Hours £	Mat'ls £	O & P £	Total £

One coat white oil-based undercoat (cont'd)

frames and linings

over 300mm girth	m2	0.24	3.60	0.44	0.61	4.65
not exceeding 150mm girth	m	0.05	0.75	0.08	0.12	0.95
150 to 300mm girth	m	0.09	1.35	0.16	0.23	1.74

skirtings and rails

over 300mm girth	m2	0.26	3.90	0.44	0.65	4.99
not exceeding 150mm girth	m	0.11	1.65	0.08	0.26	1.99
150 to 300mm girth	m	0.15	2.25	0.16	0.36	2.77

structural metalwork, general surfaces

over 300mm girth	m2	0.24	3.60	0.44	0.61	4.65
not exceeding 150mm girth	m	0.05	0.75	0.08	0.12	0.95
150 to 300mm girth	m	0.09	1.35	0.16	0.23	1.74
isolated areas not exceeding 0.5m2	nr	0.12	1.80	0.23	0.30	2.33

structural metalwork, roof truss members

over 300mm girth	m2	0.40	6.00	0.44	0.97	7.41
not exceeding 150mm girth	m	0.08	1.20	0.08	0.19	1.47
150 to 300mm girth	m	0.06	0.90	0.16	0.16	1.22
isolated areas not exceeding 0.5m2	nr	0.20	3.00	0.23	0.48	3.71

	Unit	Labour	Hours £	Mat'ls £	O & P £	Total £
radiators, panel type						
over 300mm girth	m2	0.16	2.40	0.44	0.43	3.27
not exceeding 150mm			-			
girth	m	0.05	0.75	0.08	0.12	0.95
150 to 300mm girth	m	0.09	1.35	0.16	0.23	1.74
			-			
radiators, column type			-			
over 300mm girth	m2	0.30	4.50	0.44	0.74	5.68
not exceeding 150mm						
girth	m	0.05	0.75	0.08	0.12	0.95
150 to 300mm girth	m	0.10	1.50	0.16	0.25	1.91

One coat coloured oil-based
undercoat on primed
surfaces over 300mm girth

	Unit	Labour	Hours £	Mat'ls £	O & P £	Total £
brickwork walls	m2	0.08	1.20	0.47	0.25	1.92
brickwork walls in staircase areas	m2	0.10	1.50	0.47	0.30	2.27
blockwork walls	m2	0.10	1.50	0.47	0.30	2.27
blockwork walls in staircase areas	m2	0.12	1.80	0.47	0.34	2.61
concrete walls	m2	0.11	1.65	0.47	0.32	2.44
concrete walls in staircase areas	m2	0.09	1.35	0.47	0.27	2.09
concrete ceilings	m2	0.11	1.65	0.47	0.32	2.44
concrete ceilings in staircase areas	m2	0.13	1.95	0.47	0.36	2.78
plastered walls	m2	0.06	0.90	0.47	0.21	1.58
plastered walls in staircase areas	m2	0.08	1.20	0.47	0.25	1.92
plastered ceilings	m2	0.08	1.20	0.47	0.25	1.92
plastered ceilings in staircase areas	m2	0.10	1.50	0.47	0.30	2.27

	Unit	Labour Hours	Mat'ls £	O & P £	Total £

One coat coloured oil-based undercoat (cont'd)

	Unit	Labour Hours	Mat'ls £	O & P £	Total £
embossed paper	m2	0.06	0.90	0.47	0.21	1.58
embossed paper in stair-case areas	m2	0.08	1.20	0.47	0.25	1.92
general surfaces						
over 300mm girth	m2	0.16	2.40	0.47	0.43	3.30
not exceeding 150mm girth	m	0.05	0.75	0.09	0.13	0.97
150 to 300mm girth	m	0.09	1.35	0.17	0.23	1.75
isolated areas not exceeding 0.5m2	nr	0.11	1.65	0.25	0.29	2.19
windows, screens glazed doors and the like						
panes area not exceeding 0.2m2	m2	0.33	4.95	0.33	0.79	6.07
panes area 0.1 to 0.5m2	m2	0.28	4.20	0.27	0.67	5.14
panes area 0.5 to 1m2	m2	0.23	3.45	0.23	0.55	4.23
panes area exceeding 1m2	m2	0.20	3.00	0.19	0.48	3.67
frames and linings						
over 300mm girth	m2	0.24	3.60	0.47	0.61	4.68
not exceeding 150mm girth	m	0.05	0.75	0.09	0.13	0.97
150 to 300mm girth	m	0.09	1.35	0.17	0.23	1.75
skirtings and rails						
over 300mm girth	m2	0.26	3.90	0.47	0.66	5.03
not exceeding 150mm girth	m	0.11	1.65	0.09	0.26	2.00
150 to 300mm girth	m	0.15	2.25	0.17	0.36	2.78

	Unit	Labour Hours	Mat'ls £	O & P £	Total £
structural metalwork, roof truss members					
over 300mm girth	m2	0.40	6.00	0.47	0.97	7.44
not exceeding 150mm girth	m	0.08	1.20	0.09	0.19	1.48
150 to 300mm girth	m	0.06	0.90	0.17	0.16	1.23
isolated areas not exceeding 0.5m2	nr	0.20	3.00	0.24	0.49	3.73
radiators, panel type					
over 300mm girth	m2	0.16	2.40	0.47	0.43	3.30
not exceeding 150mm girth	m	0.05	0.75	0.09	0.13	0.97
150 to 300mm girth	m	0.09	1.35	0.17	0.23	1.75
isolated areas not exceeding 0.5m2	nr	0.12	1.80	0.24	0.31	2.35
radiators, column type					
over 300mm girth	m2	0.30	4.50	0.47	0.75	5.72
not exceeding 150mm girth	m	0.05	0.75	0.09	0.13	0.97
150 to 300mm girth	m	0.10	1.50	0.17	0.25	1.92
isolated areas not exceeding 0.5m2	nr	0.15	2.25	0.24	0.37	2.86
One coat rubber paint as undercoat on primed on surfaces over 300mm girth					
brickwork walls	m2	0.20	3.00	2.31	0.80	6.11
brickwork walls in staircase areas	m2	0.22	3.30	2.31	0.84	6.45
blockwork walls	m2	0.22	3.30	2.31	0.84	6.45
blockwork walls in staircase areas	m2	0.24	3.60	2.31	0.89	6.80

	Unit	Labour Hours	Mat'ls £	O & P £	Total £

One coat rubber paint as undercoat (cont'd)

	Unit	Labour Hours	Mat'ls £	O & P £	Total £
concrete walls	m2	0.18	2.70	2.31	0.75	5.76
concrete walls in staircase areas	m2	0.20	3.00	2.31	0.80	6.11
concrete ceilings	m2	0.22	3.30	2.31	0.84	6.45
concrete ceilings in staircase areas	m2	0.24	3.60	2.31	0.89	6.80
plastered walls	m2	0.18	2.70	2.31	0.75	5.76
plastered walls in staircase areas	m2	0.20	3.00	2.31	0.80	6.11
plastered ceilings	m2	0.22	3.30	2.31	0.84	6.45
plastered ceilings in staircase areas	m2	0.24	3.60	2.31	0.89	6.80

One coat Artex sealer undercoat on surfaces over 300mm girth

	Unit	Labour Hours	Mat'ls £	O & P £	Total £
brickwork walls	m2	0.20	3.00	0.71	0.56	4.27
brickwork walls in staircase areas	m2	0.22	3.30	0.71	0.60	4.61
blockwork walls	m2	0.22	3.30	0.71	0.60	4.61
blockwork walls in staircase areas	m2	0.24	3.60	0.71	0.65	4.96
concrete walls	m2	0.18	2.70	0.71	0.51	3.92
concrete walls in staircase areas	m2	0.20	3.00	0.71	0.56	4.27
concrete ceilings	m2	0.22	3.30	0.71	0.60	4.61
concrete ceilings in staircase areas	m2	0.24	3.60	0.71	0.65	4.96
plastered walls	m2	0.18	2.70	0.71	0.51	3.92
plastered walls in staircase areas	m2	0.20	3.00	0.71	0.56	4.27

	Unit	Labour	Hours £	Mat'ls £	O & P £	Total £
plastered ceilings	m2	0.22	3.30	0.71	0.60	4.61
plastered ceilings in staircase areas	m2	0.24	3.60	0.71	0.65	4.96

One coat clear polyurethane
varnish as undercoat on
surfaces over 300mm girth

	Unit	Labour	Hours £	Mat'ls £	O & P £	Total £
general surfaces						
over 300mm girth	m2	0.16	2.40	1.76	0.62	4.78
not exceeding 150mm girth	m	0.05	0.75	0.29	0.16	1.20
150 to 300mm girth	m	0.09	1.35	0.58	0.29	2.22
isolated areas not exceeding 0.5m2	nr	0.11	1.65	0.87	0.38	2.90
windows, screens glazed doors and the like						
panes area not exceeding 0.2m2	m2	0.33	4.95	4.41	1.40	10.76
panes area 0.1 to 0.5m2	m2	0.28	4.20	4.36	1.28	9.84
panes area 0.5 to 1m2	m2	0.23	3.45	4.32	1.17	8.94
panes area exceeding 1m2	m2	0.20	3.00	4.26	1.09	8.35
frames and linings						
over 300mm girth	m2	0.24	3.60	1.76	0.80	6.16
not exceeding 150mm girth	m	0.05	0.75	0.29	0.16	1.20
150 to 300mm girth	m	0.09	1.35	0.58	0.29	2.22
skirtings and rails						
over 300mm girth	m2	0.26	3.90	1.76	0.85	6.51
not exceeding 150mm girth	m	0.11	1.65	0.29	0.29	2.23
150 to 300mm girth	m	0.15	2.25	0.58	0.42	3.25

	Unit	Labour Hours	Mat'ls £	O & P £	Total £

One coat coloured polyurethane
varnish as undercoat on
surfaces over 300mm girth

general surfaces					
over 300mm girth	m2	0.16	2.40	2.02	0.66	5.08
not exceeding 150mm girth	m	0.05	0.75	0.36	0.17	1.28
150 to 300mm girth	m	0.09	1.35	0.70	0.31	2.36
isolated areas not exceeding 0.5m2	nr	0.11	1.65	1.00	0.40	3.05

windows, screens glazed doors and the like					
panes area not exceeding 0.2m2	m2	0.33	4.95	5.15	1.52	11.62
panes area 0.1 to 0.5m2	m2	0.28	4.20	5.10	1.40	10.70
panes area 0.5 to 1m2	m2	0.23	3.45	5.05	1.28	9.78
panes area exceeding 1m2	m2	0.20	3.00	5.00	1.20	9.20

frames and linings					
over 300mm girth	m2	0.24	3.60	2.02	0.84	6.46
not exceeding 150mm girth	m	0.05	0.75	0.36	0.17	1.28
150 to 300mm girth	m	0.09	1.35	5.01	0.95	7.31

skirtings and rails					
over 300mm girth	m2	0.26	3.90	2.02	0.89	6.81
not exceeding 150mm girth	m	0.11	1.65	0.36	0.30	2.31
150 to 300mm girth	m	0.15	2.25	5.01	1.09	8.35

	Unit	Labour	Hours £	Mat'ls £	O & P £	Total £

FINISHING COATS

One coat matt white emulsion
paint on undercoated surfaces
over 300mm girth

	Unit	Labour	Hours £	Mat'ls £	O & P £	Total £
brickwork walls	m2	0.08	1.20	0.44	0.25	1.89
brickwork walls in staircase areas	m2	0.10	1.50	0.44	0.29	2.23
blockwork walls	m2	0.10	1.50	0.44	0.29	2.23
blockwork walls in staircase areas	m2	0.12	1.80	0.44	0.34	2.58
concrete walls	m2	0.07	1.05	0.44	0.22	1.71
concrete walls in staircase areas	m2	0.09	1.35	0.44	0.27	2.06
concrete ceilings	m2	0.11	1.65	0.44	0.31	2.40
concrete ceilings in staircase areas	m2	0.03	0.45	0.44	0.13	1.02
plastered walls	m2	0.06	0.90	0.44	0.20	1.54
plastered walls in staircase areas	m2	0.08	1.20	0.44	0.25	1.89
plastered ceilings	m2	0.08	1.20	0.44	0.25	1.89
plastered ceilings in staircase areas	m2	0.10	1.50	0.44	0.29	2.23
embossed paper	m2	0.06	0.90	0.44	0.20	1.54
embossed paper in staircase areas	m2	0.08	1.20	0.44	0.25	1.89

One coat matt coloured
emulsion paint on undercoated
surfaces over 300mm girth

	Unit	Labour	Hours £	Mat'ls £	O & P £	Total £
brickwork walls	m2	0.08	1.20	0.44	0.25	1.89
brickwork walls in staircase areas	m2	0.10	1.50	0.44	0.29	2.23

	Unit	Labour	Hours £	Mat'ls £	O & P £	Total £
One coat matt coloured emulsion paint (cont'd)						
blockwork walls	m2	0.10	1.50	0.44	0.29	2.23
blockwork walls in staircase areas	m2	0.12	1.80	0.44	0.34	2.58
concrete walls	m2	0.07	1.05	0.44	0.22	1.71
concrete walls in staircase areas	m2	0.09	1.35	0.44	0.27	2.06
concrete ceilings	m2	0.11	1.65	0.44	0.31	2.40
concrete ceilings in staircase areas	m2	0.03	0.45	0.44	0.13	1.02
plastered walls	m2	0.06	0.90	0.44	0.20	1.54
plastered walls in staircase areas	m2	0.08	1.20	0.44	0.25	1.89
plastered ceilings	m2	0.08	1.20	0.44	0.25	1.89
plastered ceilings in staircase areas	m2	0.10	1.50	0.44	0.29	2.23
embossed paper	m2	0.06	0.90	0.44	0.20	1.54
embossed paper in staircase areas	m2	0.08	1.20	0.44	0.25	1.89
One coat silk white emulsion paint on undercoated surfaces over 300mm girth						
brickwork walls	m2	0.08	1.20	0.48	0.25	1.93
brickwork walls in staircase areas	m2	0.10	1.50	0.48	0.30	2.28
blockwork walls	m2	0.10	1.50	0.48	0.30	2.28
blockwork walls in staircase areas	m2	0.12	1.80	0.48	0.34	2.62
concrete walls	m2	0.07	1.05	0.48	0.23	1.76
concrete walls in staircase areas	m2	0.09	1.35	0.48	0.27	2.10

	Unit	Labour Hours	Mat'ls £	O & P £	Total £
concrete ceilings	m2	0.11	1.65	0.48	0.32	2.45
concrete ceilings in stair-case areas	m2	0.03	0.45	0.48	0.14	1.07
plastered walls	m2	0.06	0.90	0.48	0.21	1.59
plastered walls in staircase areas	m2	0.08	1.20	0.48	0.25	1.93
plastered ceilings	m2	0.08	1.20	0.48	0.25	1.93
plastered ceilings in staircase areas	m2	0.10	1.50	0.48	0.30	2.28
embossed paper	m2	0.06	0.90	0.48	0.21	1.59
embossed paper in stair-case areas	m2	0.08	1.20	0.48	0.25	1.93

One coat silk coloured
emulsion paint on undercoated
surfaces over 300mm girth

	Unit	Labour Hours	Mat'ls £	O & P £	Total £
brickwork walls	m2	0.08	1.20	0.54	0.26	2.00
brickwork walls in staircase areas	m2	0.10	1.50	0.54	0.31	2.35
blockwork walls	m2	0.10	1.50	0.54	0.31	2.35
blockwork walls in staircase areas	m2	0.12	1.80	0.54	0.35	2.69
concrete walls	m2	0.07	1.05	0.54	0.24	1.83
concrete walls in staircase areas	m2	0.09	1.35	0.54	0.28	2.17
concrete ceilings	m2	0.11	1.65	0.54	0.33	2.52
concrete ceilings in stair-case areas	m2	0.03	0.45	0.54	0.15	1.14
plastered walls	m2	0.06	0.90	0.54	0.22	1.66
plastered walls in staircase areas	m2	0.08	1.20	0.54	0.26	2.00
plastered ceilings	m2	0.08	1.20	0.54	0.26	2.00
plastered ceilings in staircase areas	m2	0.10	1.50	0.54	0.31	2.35

	Unit	Labour	Hours £	Mat'ls £	O & P £	Total £

One coat silk coloured emulsion paint (cont'd)

embossed paper	m2	0.06	0.90	0.54	0.22	1.66
embossed paper in stair-case areas	m2	0.08	1.20	0.54	0.26	2.00

One coat white oil-based gloss finishing coat on undercoated surfaces over 300mm girth

brickwork walls	m2	0.08	1.20	0.54	0.26	2.00
brickwork walls in staircase areas	m2	0.10	1.50	0.54	0.31	2.35
blockwork walls	m2	0.10	1.50	0.54	0.31	2.35
blockwork walls in staircase areas	m2	0.12	1.80	0.54	0.35	2.69
concrete walls	m2	0.11	1.65	0.54	0.33	2.52
concrete walls in staircase areas	m2	0.09	1.35	0.54	0.28	2.17
concrete ceilings	m2	0.11	1.65	0.54	0.33	2.52
concrete ceilings in stair-case areas	m2	0.13	1.95	0.54	0.37	2.86
plastered walls	m2	0.06	0.90	0.54	0.22	1.66
plastered walls in staircase areas	m2	0.08	1.20	0.54	0.26	2.00
plastered ceilings	m2	0.08	1.20	0.54	0.26	2.00
plastered ceilings in staircase areas	m2	0.10	1.50	0.54	0.31	2.35
embossed paper	m2	0.06	0.90	0.54	0.22	1.66
embossed paper in stair-case areas	m2	0.08	1.20	0.54	0.26	2.00

	Unit	Labour	Hours £	Mat'ls £	O & P £	Total £
general surfaces						
over 300mm girth	m2	0.16	2.40	0.54	0.44	3.38
not exceeding 150mm girth	m	0.05	0.75	0.09	0.13	0.97
150 to 300mm girth	m	0.09	1.35	0.20	0.23	1.78
isolated areas not exceeding 0.5m2	nr	0.11	1.65	0.28	0.29	2.22
windows, screens glazed doors and the like						
panes area not exceeding 0.2m2	m2	0.33	4.95	0.39	0.80	6.14
panes area 0.1 to 0.5m2	m2	0.28	4.20	0.35	0.68	5.23
panes area 0.5 to 1m2	m2	0.23	3.45	0.32	0.57	4.34
panes area exceeding 1m2	m2	0.20	3.00	0.26	0.49	3.75
frames and linings						
over 300mm girth	m2	0.24	3.60	0.54	0.62	4.76
not exceeding 150mm girth	m	0.05	0.75	0.09	0.13	0.97
150 to 300mm girth	m	0.09	1.35	0.20	0.23	1.78
skirtings and rails						
over 300mm girth	m2	0.26	3.90	0.54	0.67	5.11
not exceeding 150mm girth	m	0.11	1.65	0.09	0.26	2.00
150 to 300mm girth	m	0.15	2.25	0.20	0.37	2.82
structural metalwork, general surfaces						
over 300mm girth	m2	0.24	3.60	0.54	0.62	4.76
not exceeding 150mm girth	m	0.05	0.75	0.09	0.13	0.97
150 to 300mm girth	m	0.09	1.35	0.20	0.23	1.78
isolated areas not exceeding 0.5m2	nr	0.12	1.80	0.28	0.31	2.39

	Unit	Labour	Hours £	Mat'ls £	O & P £	Total £

One coat white oil-based finishing coat (cont'd)

structural metalwork, roof truss members

	Unit	Labour	Hours £	Mat'ls £	O & P £	Total £
over 300mm girth	m2	0.40	6.00	0.54	0.98	7.52
not exceeding 150mm girth	m	0.08	1.20	0.09	0.19	1.48
150 to 300mm girth	m	0.06	0.90	0.20	0.17	1.27
isolated areas not exceeding 0.5m2	nr	0.20	3.00	0.28	0.49	3.77

radiators, panel type

	Unit	Labour	Hours £	Mat'ls £	O & P £	Total £
over 300mm girth	m2	0.16	2.40	0.54	0.44	3.38
not exceeding 150mm girth	m	0.05	0.75	0.09	0.13	0.97
150 to 300mm girth	m	0.09	1.35	0.20	0.23	1.78

radiators, column type

	Unit	Labour	Hours £	Mat'ls £	O & P £	Total £
over 300mm girth	m2	0.30	4.50	0.54	0.76	5.80
not exceeding 150mm girth	m	0.05	0.75	0.09	0.13	0.97
150 to 300mm girth	m	0.10	1.50	0.20	0.26	1.96

One coat coloured oil-based gloss finishing coat on under-coated surfaces over 300mm girth

	Unit	Labour	Hours £	Mat'ls £	O & P £	Total £
brickwork walls	m2	0.08	1.20	0.56	0.26	2.02
brickwork walls in staircase areas	m2	0.10	1.50	0.56	0.31	2.37
blockwork walls	m2	0.10	1.50	0.56	0.31	2.37
blockwork walls in staircase areas	m2	0.12	1.80	0.56	0.35	2.71

	Unit	Labour	Hours £	Mat'ls £	O & P £	Total £
concrete walls	m2	0.11	1.65	0.56	0.33	2.54
concrete walls in staircase areas	m2	0.09	1.35	0.56	0.29	2.20
concrete ceilings	m2	0.11	1.65	0.56	0.33	2.54
concrete ceilings in staircase areas	m2	0.13	1.95	0.56	0.38	2.89
plastered walls	m2	0.06	0.90	0.56	0.22	1.68
plastered walls in staircase areas	m2	0.08	1.20	0.56	0.26	2.02
plastered ceilings	m2	0.08	1.20	0.56	0.26	2.02
plastered ceilings in staircase areas	m2	0.10	1.50	0.56	0.31	2.37
embossed paper	m2	0.06	0.90	0.56	0.22	1.68
embossed paper in staircase areas	m2	0.08	1.20	0.56	0.26	2.02
general surfaces						
over 300mm girth	m2	0.16	2.40	0.56	0.44	3.40
not exceeding 150mm girth	m	0.05	0.75	0.10	0.13	0.98
150 to 300mm girth	m	0.09	1.35	0.21	0.23	1.79
isolated areas not exceeding 0.5m2	nr	0.11	1.65	0.30	0.29	2.24
windows, screens glazed doors and the like						
panes area not exceeding 0.2m2	m2	0.33	4.95	0.41	0.80	6.16
panes area 0.1 to 0.5m2	m2	0.28	4.20	0.37	0.69	5.26
panes area 0.5 to 1m2	m2	0.23	3.45	0.33	0.57	4.35
panes area exceeding 1m2	m2	0.20	3.00	0.29	0.49	3.78

	Unit	Labour Hours	£	Mat'ls £	O & P £	Total £

One coat coloured oil-based finishing coat (cont'd)

frames and linings

over 300mm girth	m2	0.24	3.60	0.56	0.62	4.78
not exceeding 150mm girth	m	0.05	0.75	0.10	0.13	0.98
150 to 300mm girth	m	0.09	1.35	0.21	0.23	1.79

skirtings and rails

over 300mm girth	m2	0.26	3.90	0.56	0.67	5.13
not exceeding 150mm girth	m	0.11	1.65	0.10	0.26	2.01
150 to 300mm girth	m	0.15	2.25	0.21	0.37	2.83

structural metalwork, general surfaces

over 300mm girth	m2	0.24	3.60	0.56	0.62	4.78
not exceeding 150mm girth	m	0.05	0.75	0.10	0.13	0.98
150 to 300mm girth	m	0.09	1.35	0.21	0.23	1.79
isolated areas not exceeding 0.5m2	nr	0.12	1.80	0.29	0.31	2.40

structural metalwork, roof truss members

over 300mm girth	m2	0.40	6.00	0.56	0.98	7.54
not exceeding 150mm girth	m	0.08	1.20	0.10	0.20	1.50
150 to 300mm girth	m	0.06	0.90	0.21	0.17	1.28
isolated areas not exceeding 0.5m2	nr	0.20	3.00	0.29	0.49	3.78

	Unit	Labour	Hours £	Mat'ls £	O & P £	Total £
radiators, panel type						
over 300mm girth	m2	0.16	2.40	0.56	0.44	3.40
not exceeding 150mm girth	m	0.05	0.75	0.10	0.13	0.98
150 to 300mm girth	m	0.09	1.35	0.21	0.23	1.79
radiators, column type						
over 300mm girth	m2	0.30	4.50	0.56	0.76	5.82
not exceeding 150mm girth	m	0.05	0.75	0.10	0.13	0.98
150 to 300mm girth	m	0.10	1.50	0.21	0.26	1.97

One coat white eggshell finishing coat on undercoated surfaces over 300mm girth

	Unit	Labour	Hours £	Mat'ls £	O & P £	Total £
brickwork walls	m2	0.08	1.20	0.46	0.25	1.91
brickwork walls in staircase areas	m2	0.10	1.50	0.46	0.29	2.25
blockwork walls	m2	0.10	1.50	0.46	0.29	2.25
blockwork walls in staircase areas	m2	0.12	1.80	0.46	0.34	2.60
concrete walls	m2	0.11	1.65	0.46	0.32	2.43
concrete walls in staircase areas	m2	0.09	1.35	0.46	0.27	2.08
concrete ceilings	m2	0.11	1.65	0.46	0.32	2.43
concrete ceilings in staircase areas	m2	0.13	1.95	0.46	0.36	2.77
plastered walls	m2	0.06	0.90	0.46	0.20	1.56
plastered walls in staircase areas	m2	0.08	1.20	0.46	0.25	1.91
plastered ceilings	m2	0.08	1.20	0.46	0.25	1.91
plastered ceilings in staircase areas	m2	0.10	1.50	0.46	0.29	2.25

	Unit	Labour	Hours £	Mat'ls £	O & P £	Total £
One coat white eggshell finishing coat (cont'd)						
embossed paper	m2	0.06	0.90	0.46	0.20	1.56
embossed paper in stair-case areas	m2	0.08	1.20	0.46	0.25	1.91
general surfaces						
over 300mm girth	m2	0.16	2.40	0.46	0.43	3.29
not exceeding 150mm girth	m	0.05	0.75	0.08	0.12	0.95
150 to 300mm girth	m	0.09	1.35	0.16	0.23	1.74
isolated areas not exceeding 0.5m2	nr	0.11	1.65	0.24	0.28	2.17
windows, screens glazed doors and the like						
panes area not exceeding 0.2m2	m2	0.33	4.95	0.35	0.80	6.10
panes area 0.1 to 0.5m2	m2	0.28	4.20	0.30	0.68	5.18
panes area 0.5 to 1m2	m2	0.23	3.45	0.26	0.56	4.27
panes area exceeding 1m2	m2	0.20	3.00	0.22	0.48	3.70
frames and linings						
over 300mm girth	m2	0.24	3.60	0.46	0.61	4.67
not exceeding 150mm girth	m	0.05	0.75	0.08	0.12	0.95
150 to 300mm girth	m	0.09	1.35	0.16	0.23	1.74
skirtings and rails						
over 300mm girth	m2	0.26	3.90	0.46	0.65	5.01
not exceeding 150mm girth	m	0.11	1.65	0.08	0.26	1.99
150 to 300mm girth	m	0.15	2.25	0.16	0.36	2.77

	Unit	Labour	Hours £	Mat'ls £	O & P £	Total £
structural metalwork, general surfaces						
over 300mm girth	m2	0.24	3.60	0.46	0.61	4.67
not exceeding 150mm girth	m	0.05	0.75	0.08	0.12	0.95
150 to 300mm girth	m	0.09	1.35	0.16	0.23	1.74
isolated areas not exceeding 0.5m2	nr	0.12	1.80	0.24	0.31	2.35
structural metalwork, roof truss members						
over 300mm girth	m2	0.40	6.00	0.46	0.97	7.43
not exceeding 150mm girth	m	0.08	1.20	0.08	0.19	1.47
150 to 300mm girth	m	0.06	0.90	0.16	0.16	1.22
isolated areas not exceeding 0.5m2	nr	0.20	3.00	0.24	0.49	3.73
radiators, panel type						
over 300mm girth	m2	0.16	2.40	0.46	0.43	3.29
not exceeding 150mm girth	m	0.05	0.75	0.08	0.12	0.95
150 to 300mm girth	m	0.09	1.35	0.16	0.23	1.74
radiators, column type						
over 300mm girth	m2	0.30	4.50	0.46	0.74	5.70
not exceeding 150mm girth	m	0.05	0.75	0.08	0.12	0.95
150 to 300mm girth	m	0.10	1.50	0.16	0.25	1.91

	Unit	Labour Hours	Hours £	Mat'ls £	O & P £	Total £

One coat coloured eggshell
finishing coat on undercoated
surfaces over 300mm girth

brickwork walls	m2	0.08	1.20	0.53	0.26	1.99
brickwork walls in staircase areas	m2	0.10	1.50	0.53	0.30	2.33
blockwork walls	m2	0.10	1.50	0.53	0.30	2.33
blockwork walls in staircase areas	m2	0.12	1.80	0.53	0.35	2.68
concrete walls	m2	0.11	1.65	0.53	0.33	2.51
concrete walls in staircase areas	m2	0.09	1.35	0.53	0.28	2.16
concrete ceilings	m2	0.11	1.65	0.53	0.33	2.51
concrete ceilings in staircase areas	m2	0.13	1.95	0.53	0.37	2.85
plastered walls	m2	0.06	0.90	0.53	0.21	1.64
plastered walls in staircase areas	m2	0.08	1.20	0.53	0.26	1.99
plastered ceilings	m2	0.08	1.20	0.53	0.26	1.99
plastered ceilings in staircase areas	m2	0.10	1.50	0.53	0.30	2.33
embossed paper	m2	0.06	0.90	0.53	0.21	1.64
embossed paper in staircase areas	m2	0.08	1.20	0.53	0.26	1.99
general surfaces						
over 300mm girth	m2	0.16	2.40	0.53	0.44	3.37
not exceeding 150mm girth	m	0.05	0.75	0.09	0.13	0.97
150 to 300mm girth	m	0.09	1.35	0.18	0.23	1.76
isolated areas not exceeding 0.5m2	nr	0.11	1.65	0.27	0.29	2.21

	Unit	Labour Hours	Mat'ls £	O & P £	Total £
windows, screens glazed doors and the like					
panes area not exceeding 0.2m2	m2	0.33 4.95	0.41	0.80	6.16
panes area 0.1 to 0.5m2	m2	0.28 4.20	0.37	0.69	5.26
panes area 0.5 to 1m2	m2	0.23 3.45	0.33	0.57	4.35
panes area exceeding 1m2	m2	0.20 3.00	0.29	0.49	3.78
frames and linings					
over 300mm girth	m2	0.24 3.60	0.53	0.62	4.75
not exceeding 150mm girth	m	0.05 0.75	0.09	0.13	0.97
150 to 300mm girth	m	0.09 1.35	0.18	0.23	1.76
skirtings and rails					
over 300mm girth	m2	0.26 3.90	0.53	0.66	5.09
not exceeding 150mm girth	m	0.11 1.65	0.09	0.26	2.00
150 to 300mm girth	m	0.15 2.25	0.18	0.36	2.79
structural metalwork, general surfaces					
over 300mm girth	m2	0.24 3.60	0.53	0.62	4.75
not exceeding 150mm girth	m	0.05 0.75	0.09	0.13	0.97
150 to 300mm girth	m	0.09 1.35	0.18	0.23	1.76
isolated areas not exceeding 0.5m2	nr	0.12 1.80	0.27	0.31	2.38
structural metalwork, roof truss members					
over 300mm girth	m2	0.40 6.00	0.47	0.97	7.44
not exceeding 150mm girth	m	0.08 1.20	0.08	0.19	1.47
150 to 300mm girth	m	0.06 0.90	0.15	0.16	1.21

	Unit	Labour Hours	Mat'ls £	O & P £	Total £

One coat coloured eggshell finishing coat (cont'd)

	Unit	Labour Hours	Mat'ls £	O & P £	Total £
isolated areas not exceeding 0.5m2	nr	0.20	3.00	0.24	0.49	3.73
radiators, panel type over 300mm girth	m2	0.16	2.40	0.47	0.43	3.30
not exceeding 150mm girth	m	0.05	0.75	0.08	0.12	0.95
150 to 300mm girth	m	0.09	1.35	0.15	0.23	1.73
radiators, column type over 300mm girth	m2	0.30	4.50	0.47	0.75	5.72
not exceeding 150mm girth	m	0.05	0.75	0.08	0.12	0.95
150 to 300mm girth	m	0.10	1.50	0.15	0.25	1.90

One coat standard coloured rubber finishing coat on undercoated surfaces over 300mm girth

	Unit	Labour Hours	Mat'ls £	O & P £	Total £
brickwork walls	m2	0.20	3.00	3.47	0.97	7.44
brickwork walls in staircase areas	m2	0.22	3.30	3.47	1.02	7.79
blockwork walls	m2	0.22	3.30	3.47	1.02	7.79
blockwork walls in staircase areas	m2	0.24	3.60	3.47	1.06	8.13
concrete walls	m2	0.18	2.70	3.47	0.93	7.10
concrete walls in staircase areas	m2	0.20	3.00	3.47	0.97	7.44
concrete ceilings	m2	0.22	3.30	3.47	1.02	7.79
concrete ceilings in staircase areas	m2	0.24	3.60	3.47	1.06	8.13

	Unit	Labour	Hours £	Mat'ls £	O & P £	Total £
plastered walls	m2	0.18	2.70	3.47	0.93	7.10
plastered walls in staircase areas	m2	0.20	3.00	3.47	0.97	7.44
plastered ceilings	m2	0.22	3.30	3.47	1.02	7.79
plastered ceilings in staircase areas	m2	0.24	3.60	3.47	1.06	8.13

One coat rich coloured
rubber finishing coat on
undercoated surfaces over
300mm girth

	Unit	Labour	Hours £	Mat'ls £	O & P £	Total £
brickwork walls	m2	0.20	3.00	4.14	1.07	8.21
brickwork walls in staircase areas	m2	0.22	3.30	4.14	1.12	8.56
blockwork walls	m2	0.22	3.30	4.14	1.12	8.56
blockwork walls in staircase areas	m2	0.24	3.60	4.14	1.16	8.90
concrete walls	m2	0.18	2.70	4.14	1.03	7.87
concrete walls in staircase areas	m2	0.20	3.00	4.14	1.07	8.21
concrete ceilings	m2	0.22	3.30	4.14	1.12	8.56
concrete ceilings in staircase areas	m2	0.24	3.60	4.14	1.16	8.90
plastered walls	m2	0.18	2.70	4.14	1.03	7.87
plastered walls in staircase areas	m2	0.20	3.00	4.14	1.07	8.21
plastered ceilings	m2	0.22	3.30	4.14	1.12	8.56
plastered ceilings in staircase areas	m2	0.24	3.60	4.14	1.16	8.90

	Unit	Labour	Hours £	Mat'ls £	O & P £	Total £

One coat Artex AX finishing
coat on undercoated surfaces
over 300mm girth

brickwork walls	m2	0.20	3.00	0.72	0.56	4.28
brickwork walls in staircase areas	m2	0.22	3.30	0.72	0.60	4.62
blockwork walls	m2	0.22	3.30	0.72	0.60	4.62
blockwork walls in staircase areas	m2	0.24	3.60	0.72	0.65	4.97
concrete walls	m2	0.18	2.70	0.72	0.51	3.93
concrete walls in staircase areas	m2	0.20	3.00	0.72	0.56	4.28
concrete ceilings	m2	0.22	3.30	0.72	0.60	4.62
concrete ceilings in stair-case areas	m2	0.24	3.60	0.72	0.65	4.97
plastered walls	m2	0.18	2.70	0.72	0.51	3.93
plastered walls in staircase areas	m2	0.20	3.00	0.72	0.56	4.28
plastered ceilings	m2	0.22	3.30	0.72	0.60	4.62
plastered ceilings in staircase areas	m2	0.24	3.60	0.72	0.65	4.97

One coat clear polyurethane
varnish as undercoat on
surfaces over 300mm girth

general surfaces						
over 300mm girth	m2	0.16	2.40	1.77	0.63	4.80
not exceeding 150mm girth	m	0.05	0.75	0.30	0.16	1.21
150 to 300mm girth	m	0.09	1.35	0.58	0.29	2.22
isolated areas not exceeding 0.5m2	nr	0.11	1.65	0.87	0.38	2.90

	Unit	Labour Hours	Mat'ls £	O & P £	Total £
			£		
windows, screens glazed doors and the like					
panes area not exceeding 0.2m2	m2	0.33	4.95	4.39	1.40	10.74
panes area 0.1 to 0.5m2	m2	0.28	4.20	4.35	1.28	9.83
panes area 0.5 to 1m2	m2	0.23	3.45	4.31	1.16	8.92
panes area exceeding 1m2	m2	0.20	3.00	4.27	1.09	8.36
frames and linings					
over 300mm girth	m2	0.24	3.60	1.77	0.81	6.18
not exceeding 150mm girth	m	0.05	0.75	0.30	0.16	1.21
150 to 300mm girth	m	0.09	1.35	0.58	0.29	2.22
skirtings and rails					
over 300mm girth	m2	0.26	3.90	1.77	0.85	6.52
not exceeding 150mm girth	m	0.11	1.65	0.30	0.29	2.24
150 to 300mm girth	m	0.15	2.25	0.58	0.42	3.25

One coat coloured polyurethane varnish as finishing coat on undercoated surfaces over 300mm girth

	Unit	Labour Hours	Mat'ls £	O & P £	Total £
general surfaces					
over 300mm girth	m2	0.16	2.40	2.00	0.66	5.06
not exceeding 150mm girth	m	0.05	0.75	0.34	0.16	1.25
150 to 300mm girth	m	0.09	1.35	0.67	0.30	2.32
isolated areas not exceeding 0.5m2	nr	0.11	1.65	1.00	0.40	3.05

	Unit	Labour Hours	Mat'ls £	O & P £	Total £

One coat coloured polyurethane finishing coat (cont'd)

windows, screens glazed
doors and the like
  panes area not exceeding

	Unit	Labour	Hours £	Mat'ls £	O & P £	Total £
0.2m2	m2	0.33	4.95	5.15	1.52	11.62
panes area 0.1 to 0.5m2	m2	0.28	4.20	5.11	1.40	10.71
panes area 0.5 to 1m2	m2	0.23	3.45	5.07	1.28	9.80
panes area exceeding 1m2	m2	0.20	3.00	5.03	1.20	9.23
frames and linings						
over 300mm girth	m2	0.24	3.60	2.00	0.84	6.44
not exceeding 150mm						
girth	m	0.05	0.75	0.34	0.16	1.25
150 to 300mm girth	m	0.09	1.35	0.67	0.30	2.32
skirtings and rails						
over 300mm girth	m2	0.26	3.90	2.00	0.89	6.79
not exceeding 150mm						
girth	m	0.11	1.65	0.34	0.30	2.29
150 to 300mm girth	m	0.15	2.25	0.67	0.44	3.36

	Unit	Labour Hours	Mat'ls £	O & P £	Total £

WALLPAPERING

Strip off one layer of paper,
stop cracks and rub down

woodchip paper

	Unit	Labour Hours	Mat'ls £	O & P £	Total £
walls	m2	0.10	1.50	0.05	0.23	1.78
walls in staircase areas	m2	0.14	2.10	0.05	0.32	2.47
ceilings	m2	0.16	2.40	0.05	0.37	2.82
ceilings in staircase areas	m2	0.20	3.00	0.05	0.46	3.51

vinyl paper

walls	m2	0.12	1.80	0.05	0.28	2.13
walls in staircase areas	m2	0.16	2.40	0.05	0.37	2.82
ceilings	m2	0.18	2.70	0.05	0.41	3.16
ceilings in staircase areas	m2	0.22	3.30	0.05	0.50	3.85

standard patterned paper

walls	m2	0.10	1.50	0.05	0.23	1.78
walls in staircase areas	m2	0.14	2.10	0.05	0.32	2.47
ceilings	m2	0.16	2.40	0.05	0.37	2.82
ceilings in staircase areas	m2	0.20	3.00	0.05	0.46	3.51

embossed paper

walls	m2	0.12	1.80	0.05	0.28	2.13
walls in staircase areas	m2	0.16	2.40	0.05	0.37	2.82
ceilings	m2	0.18	2.70	0.05	0.41	3.16
ceilings in staircase areas	m2	0.22	3.30	0.05	0.50	3.85

Lincrusta paper

walls	m2	0.14	2.10	0.05	0.32	2.47
walls in staircase areas	m2	0.18	2.70	0.05	0.41	3.16
ceilings	m2	0.20	3.00	0.05	0.46	3.51
ceilings in staircase areas	m2	0.24	3.60	0.05	0.55	4.20

	Unit	Labour Hours £	Mat'ls £	O & P £	Total £
Strip off one layer of paper, (cont'd)					
Anaglypta paper					
walls	m2	0.20	3.00	0.05	0.46	3.51
walls in staircase areas	m2	0.24	3.60	0.05	0.55	4.20
ceilings	m2	0.26	3.90	0.05	0.59	4.54
ceilings in staircase areas	m2	0.30	4.50	0.05	0.68	5.23
Strip off two layers of paper, stop cracks and rub down					
woodchip paper					
walls	m2	0.22	3.30	0.05	0.50	3.85
walls in staircase areas	m2	0.26	3.90	0.05	0.59	4.54
ceilings	m2	0.28	4.20	0.05	0.64	4.89
ceilings in staircase areas	m2	0.32	4.80	0.05	0.73	5.58
vinyl paper					
walls	m2	0.20	3.00	0.05	0.46	3.51
walls in staircase areas	m2	0.24	3.60	0.05	0.55	4.20
ceilings	m2	0.26	3.90	0.05	0.59	4.54
ceilings in staircase areas	m2	0.30	4.50	0.05	0.68	5.23
standard patterned paper					
walls	m2	0.22	3.30	0.05	0.50	3.85
walls in staircase areas	m2	0.26	3.90	0.05	0.59	4.54
ceilings	m2	0.28	4.20	0.05	0.64	4.89
ceilings in staircase areas	m2	0.32	4.80	0.05	0.73	5.58
embossed paper					
walls	m2	0.20	3.00	0.05	0.46	3.51
walls in staircase areas	m2	0.24	3.60	0.05	0.55	4.20
ceilings	m2	0.26	3.90	0.05	0.59	4.54
ceilings in staircase areas	m2	0.30	4.50	0.05	0.68	5.23

	Unit	Labour Hours	Mat'ls £	O & P £	Total £
Prepare, size, apply adhesive, supply and hang paper to plastered walls, butt jointed					
lining paper					
£1.50 per roll	m2	0.25	3.75	0.36	0.62	4.73
£2.00 per roll	m2	0.25	3.75	0.48	0.63	4.86
£2.50 per roll	m2	0.25	3.75	0.59	0.65	4.99
woodchip paper					
£2.50 per roll	m2	0.30	4.50	0.59	0.76	5.85
£4.00 per roll	m2	0.30	4.50	0.94	0.82	6.26
£5.00 per roll	m2	0.30	4.50	1.18	0.85	6.53
vinyl paper					
£4.00 per roll	m2	0.30	4.50	0.94	0.82	6.26
£5.00 per roll	m2	0.30	4.50	1.18	0.85	6.53
£6.00 per roll	m2	0.30	4.50	1.41	0.89	6.80
washable paper					
£5.00 per roll	m2	0.30	4.50	1.18	0.85	6.53
£6.00 per roll	m2	0.30	4.50	1.41	0.89	6.80
£7.00 per roll	m2	0.30	4.50	1.64	0.92	7.06
hessian paper					
£7.00 per roll	m2	0.50	7.50	1.64	1.37	10.51
£8.00 per roll	m2	0.50	7.50	8.80	2.45	18.75
£9.00 per roll	m2	0.50	7.50	9.90	2.61	20.01
suede paper					
£9.00 per m2	m2	0.50	7.50	9.90	2.61	20.01
£10.00 per m2	m2	0.50	7.50	11.00	2.78	21.28
£11.00 per m2	m2	0.50	7.50	12.10	2.94	22.54

	Unit	Labour Hours	Mat'ls £	O & P £	Total £

Prepare, size, apply adhesive,
supply and hang paper to
plastered ceilings, butt jointed

	Unit	Labour Hours	Mat'ls £	O & P £	Total £
lining paper					
£1.50 per roll	m2	0.30	4.50	0.36	0.73	5.59
£2.00 per roll	m2	0.30	4.50	0.48	0.75	5.73
£2.50 per roll	m2	0.30	4.50	0.59	0.76	5.85
woodchip paper					
£2.50 per roll	m2	0.35	5.25	0.59	0.88	6.72
£4.00 per roll	m2	0.35	5.25	0.94	0.93	7.12
£5.00 per roll	m2	0.35	5.25	1.18	0.96	7.39
vinyl paper					
£4.00 per roll	m2	0.35	5.25	0.94	0.93	7.12
£5.00 per roll	m2	0.35	5.25	1.18	0.96	7.39
£6.00 per roll	m2	0.35	5.25	1.41	1.00	7.66
washable paper					
£5.00 per roll	m2	0.35	5.25	1.18	0.96	7.39
£6.00 per roll	m2	0.35	5.25	1.41	1.00	7.66
£7.00 per roll	m2	0.35	5.25	1.64	1.03	7.92

	Unit	Labour Hours	Mat'ls £	O & P £	Total £

EXTERNAL WORK

The unit rates in this section
include all the necessary
preparatory work involved
in treating existing surfaces
to receive new paintwork.
This includes cutting out and
filling cracks, knotting and
stopping where necessary
and a minimum allowance
for patch priming.

The rates are based upon
brush application and working
in normal conditions on
surfaces that could be described
as in 'average condition'.

The following adjustments
apply where necessary:

working at heights over 3m
   add 15% to labour

working on surfaces in poor
condition
   add 10% to labour and
   materials

working in cramped conditions
   add 15% to labour

spraying
   deduct 30% of labour and
   add 15% to materials

	Unit	Labour £	Hours £	Mat'ls £	O & P £	Total £

PREPARATORY WORK

Wash down oil—painted wood surfaces and rub down

general surfaces

over 300mm girth	m2	0.32	4.80	-	0.72	5.52
not exceeding 300mm girth	m	0.11	1.65	-	0.25	1.90
150 to 300mm girth	m	0.10	1.49	-	0.22	1.71
isolated areas not exceeding 0.5m2	nr	0.16	2.37	-	0.36	2.73

windows, screens glazed doors and the like

panes area not exceeding 0.2m2	m2	0.42	6.30	-	0.95	7.25
panes area 0.1 to 0.5m2	m2	0.39	5.85	-	0.88	6.73
panes area 0.5 to 1m2	m2	0.36	5.40	-	0.81	6.21
panes area exceeding 1m2	m2	0.32	4.80	-	0.72	5.52

Wash down oil—painted metal surfaces and rub down

general surfaces

over 300mm girth	m2	0.32	4.80	-	0.72	5.52
not exceeding 300mm girth	m	0.11	1.65	-	0.25	1.90
150 to 300mm girth	m	0.10	1.49	-	0.22	1.71
isolated areas not exceeding 0.5m2	nr	0.16	2.37	-	0.36	2.73

	Unit	Labour Hours	Mat'ls £	O & P £	Total £
		£			
windows, screens glazed doors and the like					
panes area not exceeding 0.2m2	m2	0.42	6.30	-	0.95	7.25
panes area 0.1 to 0.5m2	m2	0.39	5.85	-	0.88	6.73
panes area 0.5 to 1m2	m2	0.36	5.40	-	0.81	6.21
panes area exceeding 1m2	m2	0.32	4.80	-	0.72	5.52
structural metalwork, general surfaces					
over 300mm girth	m2	0.39	5.85	-	0.88	6.73
not exceeding 150mm girth	m	0.07	1.05	-	0.16	1.21
150 to 300mm girth	m	0.13	1.95	-	0.29	2.24
isolated areas not exceeding 0.5m2	nr	0.19	2.85	-	0.43	3.28

Wash down oil—painted wood surfaces, remove paint with chemical stripper and rub down

	Unit	Labour Hours	Mat'ls £	O & P £	Total £
general surfaces					
over 300mm girth	m2	0.82	12.30	2.58	2.23	17.11
not exceeding 300mm girth	m	0.31	4.65	0.96	0.84	6.45
150 to 300mm girth	m	0.27	4.05	0.65	0.71	5.41
isolated areas not exceeding 0.5m2	nr	0.41	6.15	1.30	1.12	8.57

	Unit	Labour Hours	Mat'ls £	O & P £	Total £

Wash down oil—painted wood surfaces (cont'd)

windows, screens glazed
doors and the like
  panes area not exceeding

0.2m2	m2	1.00	15.00	2.15	2.57	19.72
panes area 0.1 to 0.5m2	m2	0.95	14.25	2.05	2.45	18.75
panes area 0.5 to 1m2	m2	0.90	13.50	1.94	2.32	17.76
panes area exceeding 1m2	m2	0.85	12.75	1.82	2.19	16.76

Wash down oil—painted metal surfaces, remove paint with chemical stripper and rub down

general surfaces

over 300mm girth	m2	0.82	12.30	2.58	2.23	17.11
not exceeding 300mm girth	m	0.31	4.65	1.00	0.85	6.50
150 to 300mm girth	m	0.27	4.05	0.66	0.71	5.42
isolated areas not exceeding 0.5m2	nr	0.41	6.15	1.30	1.12	8.57

windows, screens glazed
doors and the like
  panes area not exceeding

0.2m2	m2	1.00	15.00	2.15	2.57	19.72
panes area 0.1 to 0.5m2	m2	0.95	14.25	2.05	2.45	18.75
panes area 0.5 to 1m2	m2	0.90	13.50	1.94	2.32	17.76
panes area exceeding 1m2	m2	0.85	12.75	1.82	2.19	16.76

	Unit	Labour	Hours £	Mat'ls £	O & P £	Total £
structural metalwork, general surfaces						
over 300mm girth	m2	0.82	12.30	2.58	2.23	17.11
not exceeding 150mm girth	m	0.32	4.80	1.00	0.87	6.67
150 to 300mm girth	m	0.42	6.30	0.66	1.04	8.00
isolated areas not exceeding 0.5m2	nr	0.41	6.15	1.30	1.12	8.57

Burn off paint from wood surfaces and rub down

	Unit	Labour	Hours £	Mat'ls £	O & P £	Total £
general surfaces						
over 300mm girth	m2	1.10	16.50	-	2.48	18.98
not exceeding 300mm girth	m	0.31	4.65	-	0.70	5.35
150 to 300mm girth	m	0.41	6.15	-	0.92	7.07
isolated areas not exceeding 0.5m2	nr	0.55	8.25	-	1.24	9.49
windows, screens glazed doors and the like						
panes area not exceeding 0.2m2	m2	1.72	25.80	-	3.87	29.67
panes area 0.1 to 0.5m2	m2	1.62	24.30	-	3.65	27.95
panes area 0.5 to 1m2	m2	1.52	22.80	-	3.42	26.22
panes area exceeding 1m2	m2	1.42	21.30	-	3.20	24.50

Burn off paint from metal surfaces and rub down

	Unit	Labour	Hours £	Mat'ls £	O & P £	Total £
general surfaces						
over 300mm girth	m2	1.10	16.50	-	2.48	18.98
not exceeding 300mm girth	m	0.31	4.65	-	0.70	5.35

	Unit	Labour	Hours £	Mat'ls £	O & P £	Total £
Burn off paint from metal surfaces (cont'd)						
150 to 300mm girth	m	0.41	6.15	-	0.92	7.07
isolated areas not exceeding 0.5m2	nr	0.55	8.25	-	1.24	9.49
windows, screens glazed doors and the like panes area not exceeding 0.2m2	m2	1.72	25.80	-	3.87	29.67
panes area 0.1 to 0.5m2	m2	1.62	24.30	-	3.65	27.95
panes area 0.5 to 1m2	m2	1.52	22.80	-	3.42	26.22
panes area exceeding 1m2	m2	1.42	21.30	-	3.20	24.50
structural metalwork, general surfaces over 300mm girth	m2	0.82	12.30	-	1.85	14.15
not exceeding 150mm girth	m	0.13	1.95	-	0.29	2.24
150 to 300mm girth	m	0.42	6.30	-	0.95	7.25
isolated areas not exceeding 0.5m2	nr	0.41	6.15	-	0.92	7.07
PRIMERS						
One coat wood primer on wood						
general surfaces over 300mm girth	m2	0.22	3.30	0.79	0.61	4.70
not exceeding 150mm girth	m	0.07	1.05	0.14	0.18	1.37
150 to 300mm girth	m	0.11	1.65	0.32	0.30	2.27
isolated areas not exceeding 0.5m2	nr	0.12	1.80	0.40	0.33	2.53

	Unit	Labour Hours	Mat'ls £	O & P £	Total £
windows, screens glazed doors and the like					
panes area not exceeding 0.2m2	m2	0.39	5.85	0.42	0.94	7.21
panes area 0.1 to 0.5m2	m2	0.34	5.10	0.38	0.82	6.30
panes area 0.5 to 1m2	m2	0.29	4.35	0.33	0.70	5.38
panes area exceeding 1m2	m2	0.26	3.90	0.29	0.63	4.82
frames and linings					
over 300mm girth	m2	0.30	4.50	0.79	0.79	6.08
not exceeding 150mm girth	m	0.07	1.05	0.14	0.18	1.37
150 to 300mm girth	m	0.11	1.65	0.32	0.30	2.27
railings, fences and gates plain open type					
over 300mm girth	m2	0.30	4.50	0.79	0.79	6.08
not exceeding 150mm girth	m	0.07	1.05	0.14	0.18	1.37
150 to 300mm girth	m	0.11	1.65	0.32	0.30	2.27
railings, fences and gates ornamental type					
over 300mm girth	m2	0.26	3.90	0.79	0.70	5.39
not exceeding 150mm girth	m	0.06	0.90	0.14	0.16	1.20
150 to 300mm girth	m	0.10	1.50	0.32	0.27	2.09

One coat aluminium primer on wood

	Unit	Labour Hours	Mat'ls £	O & P £	Total £
general surfaces					
over 300mm girth	m2	0.22	3.30	1.44	0.71	5.45
not exceeding 150mm girth	m	0.07	1.05	0.26	0.20	1.51
150 to 300mm girth	m	0.11	1.65	0.49	0.32	2.46

	Unit	Labour Hours	Mat'ls £	O & P £	Total £

One coat aluminium primer (cont'd)

	Unit	Labour Hours	Mat'ls £	O & P £	Total £
isolated areas not exceeding 0.5m2	nr	0.12	1.80	0.72	0.38	2.90
windows, screens glazed doors and the like panes area not exceeding 0.2m2	m2	0.39	5.85	0.72	0.99	7.56
panes area 0.1 to 0.5m2	m2	0.34	5.10	0.67	0.87	6.64
panes area 0.5 to 1m2	m2	0.29	4.35	0.63	0.75	5.73
panes area exceeding 1m2	m2	0.26	3.90	0.58	0.67	5.15
frames and linings over 300mm girth	m2	0.30	4.50	1.44	0.89	6.83
not exceeding 150mm girth	m	0.07	1.05	0.26	0.20	1.51
150 to 300mm girth	m	0.11	1.65	0.49	0.32	2.46
railings, fences and gates plain open type over 300mm girth	m2	0.30	4.50	1.44	0.89	6.83
not exceeding 150mm girth	m	0.07	1.05	0.26	0.20	1.51
150 to 300mm girth	m	0.11	1.65	0.49	0.32	2.46
railings, fences and gates ornamental type over 300mm girth	m2	0.26	3.90	1.44	0.80	6.14
not exceeding 150mm girth	m	0.06	0.90	0.26	0.17	1.33
150 to 300mm girth	m	0.10	1.50	0.49	0.30	2.29

	Unit	Labour Hours	Mat'ls £	O & P £	Total £

One coat acrylic primer on wood

general surfaces

	Unit	Labour Hours	Mat'ls £	O & P £	Total £
over 300mm girth	m2	0.22	3.30	0.79	0.61	4.70
not exceeding 150mm girth	m	0.07	1.05	0.14	0.18	1.37
150 to 300mm girth	m	0.11	1.65	0.32	0.30	2.27
isolated areas not exceeding 0.5m2	nr	0.12	1.80	0.39	0.33	2.52

windows, screens glazed doors and the like

panes area not exceeding 0.2m2	m2	0.39	5.85	0.42	0.94	7.21
panes area 0.1 to 0.5m2	m2	0.34	5.10	0.38	0.82	6.30
panes area 0.5 to 1m2	m2	0.29	4.35	0.33	0.70	5.38
panes area exceeding 1m2	m2	0.26	3.90	0.29	0.63	4.82

frames and linings

over 300mm girth	m2	0.30	4.50	0.79	0.79	6.08
not exceeding 150mm girth	m	0.07	1.05	0.14	0.18	1.37
150 to 300mm girth	m	0.11	1.65	0.32	0.30	2.27

railings, fences and gates plain open type

over 300mm girth	m2	0.30	4.50	0.79	0.79	6.08
not exceeding 150mm girth	m	0.07	1.05	0.14	0.18	1.37
150 to 300mm girth	m	0.11	1.65	0.32	0.30	2.27

railings, fences and gates ornamental type

over 300mm girth	m2	0.26	3.90	0.79	0.70	5.39

	Unit	Labour Hours	Mat'ls £	O & P £	Total £

One coat acrylic primer (cont'd)

not exceeding 150mm girth	m	0.06 0.90	0.14	0.16	1.20
150 to 300mm girth	m	0.10 1.50	0.32	0.27	2.09

One coat zinc chromate primer on metal

general surfaces					
over 300mm girth	m2	0.22 3.30	0.61	0.59	4.50
not exceeding 150mm girth	m	0.07 1.05	0.11	0.17	1.33
150 to 300mm girth	m	0.11 1.65	0.21	0.28	2.14
isolated areas not exceeding 0.5m2	nr	0.12 1.80	0.31	0.32	2.43

windows, screens glazed doors and the like					
panes area not exceeding 0.2m2	m2	0.39 5.85	0.31	0.92	7.08
panes area 0.1 to 0.5m2	m2	0.34 5.10	0.26	0.80	6.16
panes area 0.5 to 1m2	m2	0.29 4.35	0.22	0.69	5.26
panes area exceeding 1m2	m2	0.26 3.90	0.18	0.61	4.69

structural metalwork, general surfaces					
over 300mm girth	m2	0.34 5.10	0.61	0.86	6.57
not exceeding 150mm girth	m	0.06 0.90	0.11	0.15	1.16
150 to 300mm girth	m	0.11 1.65	0.21	0.28	2.14
isolated areas not exceeding 0.5m2	nr	0.17 2.55	0.31	0.43	3.29

	Unit	Labour	Hours £	Mat'ls £	O & P £	Total £
eaves gutters						
over 300mm girth	m2	0.27	4.05	0.61	0.70	5.36
not exceeding 300mm girth	m	0.10	1.50	0.21	0.26	1.97
isolated areas not exceeding 0.5m2	nr	0.14	2.10	0.31	0.36	2.77
services, pipes, conduits ducting and the like						
over 300mm girth	m2	0.27	4.05	0.61	0.70	5.36
not exceeding 300mm girth	m	0.10	1.50	0.21	0.26	1.97
isolated areas not exceeding 0.5m2	nr	0.14	2.10	0.31	0.36	2.77

One coat metal red oxide primer on metal

	Unit	Labour	Hours £	Mat'ls £	O & P £	Total £
general surfaces						
over 300mm girth	m2	0.22	3.30	0.56	0.58	4.44
not exceeding 150mm girth	m	0.07	1.05	0.10	0.17	1.32
150 to 300mm girth	m	0.11	1.65	0.19	0.28	2.12
isolated areas not exceeding 0.5m2	nr	0.12	1.80	0.29	0.31	2.40
windows, screens glazed doors and the like						
panes area not exceeding 0.2m2	m2	0.39	5.85	0.28	0.92	7.05
panes area 0.1 to 0.5m2	m2	0.34	5.10	0.24	0.80	6.14
panes area 0.5 to 1m2	m2	0.29	4.35	0.20	0.68	5.23
panes area exceeding 1m2	m2	0.26	3.90	0.16	0.61	4.67

	Unit	Labour Hours £	Mat'ls £	O & P £	Total £

One coat metal red oxide
(cont'd)

strructural metalwork, general
surfaces

over 300mm girth	m2	0.34	5.10	0.56	0.85	6.51
not exceeding 150mm girth	m	0.06	0.90	0.10	0.15	1.15
150 to 300mm girth	m	0.11	1.65	0.19	0.28	2.12
isolated areas not exceeding 0.5m2	nr	0.17	2.55	0.29	0.43	3.27

eaves gutters

over 300mm girth	m2	0.27	4.05	0.56	0.69	5.30
not exceeding 300mm girth	m	0.10	1.50	0.19	0.25	1.94
isolated areas not exceeding 0.5m2	nr	0.14	2.10	0.29	0.36	2.75

services, pipes, conduits
ducting and the like

over 300mm girth	m2	0.27	4.05	0.56	0.69	5.30
not exceeding 300mm girth	m	0.10	1.50	0.19	0.25	1.94
isolated areas not exceeding 0.5m2	nr	0.14	2.10	0.29	0.36	2.75

UNDERCOATS

One coat white oil-based
undercoat on primed
surfaces over 300mm girth

general surfaces

over 300mm girth	m2	0.20	3.00	0.44	0.52	3.96

	Unit	Labour	Hours £	Mat'ls £	O & P £	Total £
not exceeding 150mm girth	m	0.06	0.90	0.08	0.15	1.13
150 to 300mm girth	m	0.10	1.50	0.16	0.25	1.91
isolated areas not exceeding 0.5m2	nr	0.11	1.65	0.24	0.28	2.17
windows, screens glazed doors and the like						
panes area not exceeding 0.2m2	m2	0.37	5.55	0.30	0.88	6.73
panes area 0.1 to 0.5m2	m2	0.32	4.80	0.26	0.76	5.82
panes area 0.5 to 1m2	m2	0.27	4.05	0.22	0.64	4.91
panes area exceeding 1m2	m2	0.24	3.60	0.18	0.57	4.35
frames and linings						
over 300mm girth	m2	0.28	4.20	0.44	0.70	5.34
not exceeding 150mm girth	m	0.06	0.90	0.08	0.15	1.13
150 to 300mm girth	m	0.10	1.50	0.16	0.25	1.91
railings, fences and gates plain open type						
over 300mm girth	m2	0.28	4.20	0.44	0.70	5.34
not exceeding 150mm girth	m	0.06	0.90	0.08	0.15	1.13
150 to 300mm girth	m	0.11	1.65	0.16	0.27	2.08
railings, fences and gates ornamental type						
over 300mm girth	m2	0.26	3.90	0.78	0.70	5.38
not exceeding 150mm girth	m	0.05	0.75	0.14	0.13	1.02
150 to 300mm girth	m	0.10	1.50	0.16	0.25	1.91

	Unit	Labour Hours	£	Mat'ls £	O & P £	Total £

One coat white oil-based undercoat (cont'd)

eaves gutters

	Unit	Labour Hours	£	Mat'ls £	O & P £	Total £
over 300mm girth	m2	0.27	4.05	0.44	0.67	5.16
not exceeding 300mm girth	m	0.10	1.50	0.16	0.25	1.91
isolated areas not exceeding 0.5m2	nr	0.14	2.10	0.22	0.35	2.67

services, pipes, conduits ducting and the like

	Unit	Labour Hours	£	Mat'ls £	O & P £	Total £
over 300mm girth	m2	0.27	4.05	0.44	0.67	5.16
not exceeding 300mm girth	m	0.10	1.50	0.16	0.25	1.91
isolated areas not exceeding 0.5m2	nr	0.14	2.10	0.22	0.35	2.67

One coat coloured oil-based undercoat on primed surfaces over 300mm girth

general surfaces

	Unit	Labour Hours	£	Mat'ls £	O & P £	Total £
over 300mm girth	m2	0.20	3.00	0.47	0.52	3.99
not exceeding 150mm girth	m	0.06	0.90	0.09	0.15	1.14
150 to 300mm girth	m	0.10	1.50	0.17	0.25	1.92
isolated areas not exceeding 0.5m2	nr	0.11	1.65	0.24	0.28	2.17

windows, screens glazed doors and the like
panes area not exceeding

	Unit	Labour Hours	£	Mat'ls £	O & P £	Total £
0.2m2	m2	0.37	5.55	0.33	0.88	6.76
panes area 0.1 to 0.5m2	m2	0.32	4.80	0.27	0.76	5.83

	Unit	Labour	Hours £	Mat'ls £	O & P £	Total £
panes area 0.5 to 1m2	m2	0.27	4.05	0.23	0.64	4.92
panes area exceeding 1m2	m2	0.24	3.60	0.19	0.57	4.36
frames and linings						
over 300mm girth	m2	0.28	4.20	0.47	0.70	5.37
not exceeding 150mm girth	m	0.06	0.90	0.09	0.15	1.14
150 to 300mm girth	m	0.10	1.50	0.17	0.25	1.92
eaves gutters						
over 300mm girth	m2	0.27	4.05	0.47	0.68	5.20
not exceeding 300mm girth	m	0.10	1.50	0.18	0.25	1.93
isolated areas not exceeding 0.5m2	nr	0.14	2.10	0.24	0.35	2.69
services, pipes, conduits ducting and the like						
over 300mm girth	m2	0.27	4.05	0.47	0.68	5.20
not exceeding 300mm girth	m	0.10	1.50	0.18	0.25	1.93
isolated areas not exceeding 0.5m2	nr	0.14	2.10	0.24	0.35	2.69
railings, fences and gates plain open type						
over 300mm girth	m2	0.28	4.20	0.44	0.70	5.34
not exceeding 150mm girth	m	0.06	0.90	0.08	0.15	1.13
150 to 300mm girth	m	0.11	1.65	0.16	0.27	2.08
railings, fences and gates ornamental type						
over 300mm girth	m2	0.26	3.90	0.77	0.70	5.37
not exceeding 150mm girth	m	0.05	0.75	0.14	0.13	1.02
150 to 300mm girth	m	0.10	1.50	0.16	0.25	1.91

	Unit	Labour Hours	Mat'ls £	O & P £	Total £

One coat masonry sealer to
surfaces over 300mm girth

brickwork walls	m2	0.12	1.80	1.30	0.47	3.57
blockwork walls	m2	0.14	2.10	1.30	0.51	3.91
concrete walls	m2	0.11	1.65	1.11	0.41	3.17
rendered walls	m2	0.12	1.80	1.11	0.44	3.35
roughcast walls	m2	0.14	2.10	1.42	0.53	4.05

One coat stabilising solution
to surfaces over 300mm girth

brickwork walls	m2	0.12	1.80	1.30	0.47	3.57
blockwork walls	m2	0.14	2.10	1.30	0.51	3.91
concrete walls	m2	0.11	1.65	1.11	0.41	3.17
rendered walls	m2	0.12	1.80	1.11	0.44	3.35
roughcast walls	m2	0.14	2.10	1.42	0.53	4.05

One coat Snowcem on primed
surfaces over 300mm girth

brickwork walls	m2	0.12	1.80	0.68	0.37	2.85
blockwork walls	m2	0.14	2.10	0.68	0.42	3.20
concrete walls	m2	0.11	1.65	0.62	0.34	2.61
rendered walls	m2	0.12	1.80	0.62	0.36	2.78
roughcast walls	m2	0.14	2.10	0.76	0.43	3.29

One coat Sandtex matt on primed
surfaces over 300mm girth

brickwork walls	m2	0.12	1.80	1.30	0.47	3.57
blockwork walls	m2	0.14	2.10	1.30	0.51	3.91
concrete walls	m2	0.11	1.65	1.11	0.41	3.17
rendered walls	m2	0.12	1.80	1.11	0.44	3.35
roughcast walls	m2	0.14	2.10	1.42	0.53	4.05

	Unit	Labour Hours	Mat'ls £	O & P £	Total £
One coat Sandtex textured finish on surfaces over 300mm girth					
brickwork walls	m2	0.12	1.80	3.74	0.83	6.37
blockwork walls	m2	0.14	2.10	3.74	0.88	6.72
concrete walls	m2	0.11	1.65	3.54	0.78	5.97
rendered walls	m2	0.12	1.80	3.54	0.80	6.14
roughcast walls	m2	0.14	2.10	3.94	0.91	6.95
One coat clear preserver on sawn wood surfaces					
over 300mm girth	m2	0.38	5.70	0.94	1.00	7.64
not exceeding 300mm girth	m	0.13	1.95	0.30	0.34	2.59
isolated areas not exceeding 0.5m2	nr	0.19	2.85	0.48	0.50	3.83
railings, fences and gates plain open type					
over 300mm girth	m2	0.30	4.50	0.94	0.82	6.26
not exceeding 150mm girth	m	0.07	1.05	0.16	0.18	1.39
150 to 300mm girth	m	0.11	1.65	0.30	0.29	2.24
railings, fences and gates ornamental type					
over 300mm girth	m2	0.26	3.90	0.94	0.73	5.57
not exceeding 150mm girth	m	0.06	0.90	0.16	0.16	1.22
150 to 300mm girth	m	0.10	1.50	0.30	0.27	2.07

	Unit	Labour Hours	Hours £	Mat'ls £	O & P £	Total £
One coat clear preserver on wrot wood surfaces						
over 300mm girth	m2	0.36	5.40	0.92	0.95	7.27
not exceeding 150mm girth	m	0.06	0.90	0.16	0.16	1.22
isolated areas not exceeding 0.5m2	nr	0.18	2.70	0.46	0.47	3.63
railings, fences and gates plain open type						
over 300mm girth	m2	0.28	4.20	0.92	0.77	5.89
not exceeding 150mm girth	m	0.06	0.90	0.16	0.16	1.22
150 to 300mm girth	m	0.10	1.50	0.30	0.27	2.07
railings, fences and gates ornamental type						
over 300mm girth	m2	0.24	3.60	0.96	0.68	5.24
not exceeding 150mm girth	m	0.05	0.75	0.29	0.16	1.20
150 to 300mm girth	m	0.09	1.35	0.47	0.27	2.09
One coat light preserver on sawn wood surfaces						
over 300mm girth	m2	0.38	5.70	0.93	0.99	7.62
not exceeding 300mm girth	m	0.13	1.95	0.30	0.34	2.59
isolated areas not exceeding 0.5m2	nr	0.19	2.85	0.47	0.50	3.82

	Unit	Labour	Hours £	Mat'ls £	O & P £	Total £
railings, fences and gates ornamental type						
over 300mm girth	m2	0.30	4.50	0.93	0.81	6.24
not exceeding 150mm girth	m	0.07	1.05	0.16	0.18	1.39
150 to 300mm girth	m	0.11	1.65	0.30	0.29	2.24
railings, fences and gates ornamental type						
over 300mm girth	m2	0.26	3.90	0.93	0.72	5.55
not exceeding 150mm girth	m	0.06	0.90	0.16	0.16	1.22
150 to 300mm girth	m	0.10	1.50	0.30	0.27	2.07

One coat light preserver on wrot wood surfaces

	Unit	Labour	Hours £	Mat'ls £	O & P £	Total £
over 300mm girth	m2	0.36	5.40	0.86	0.94	7.20
not exceeding 150mm girth	m	0.06	0.90	0.25	0.17	1.32
isolated areas not exceeding 0.5m2	nr	0.18	2.70	0.44	0.47	3.61
railings, fences and gates plain open type						
over 300mm girth	m2	0.30	4.50	0.86	0.80	6.16
not exceeding 150mm girth	m	0.07	1.05	0.14	0.18	1.37
150 to 300mm girth	m	0.11	1.65	0.29	0.29	2.23
railings, fences and gates ornamental type						
over 300mm girth	m2	0.26	3.90	0.86	0.71	5.47
not exceeding 150mm girth	m	0.06	0.90	0.14	0.16	1.20
150 to 300mm girth	m	0.10	1.50	0.29	0.27	2.06

	Unit	Labour Hours	£	Mat'ls £	O & P £	Total £
One coat dark preserver on sawn wood surfaces						
over 300mm girth	m2	0.38	5.70	0.93	0.99	7.62
not exceeding 300mm girth	m	0.13	1.95	0.30	0.34	2.59
isolated areas not exceeding 0.5m2	nr	0.19	2.85	0.47	0.50	3.82
railings, fences and gates plain open type						
over 300mm girth	m2	0.30	4.50	0.93	0.81	6.24
not exceeding 150mm girth	m	0.07	1.05	0.16	0.18	1.39
150 to 300mm girth	m	0.11	1.65	0.30	0.29	2.24
railings, fences and gates ornamental type						
over 300mm girth	m2	0.26	3.90	0.93	0.72	5.55
not exceeding 150mm girth	m	0.06	0.90	0.30	0.18	1.38
150 to 300mm girth	m	0.10	1.50	0.47	0.30	2.27
One coat dark preserver on wrot wood surfaces						
over 300mm girth	m2	0.36	5.40	0.86	0.94	7.20
not exceeding 300mm girth	m	0.12	1.80	0.27	0.31	2.38
isolated areas not exceeding 0.5m2	nr	0.18	2.70	0.45	0.47	3.62

	Unit	Labour Hours	Mat'ls £	O & P £	Total £
railings, fences and gates plain open type					
over 300mm girth	m2	0.30	4.50	0.86	0.80	6.16
not exceeding 150mm girth	m	0.07	1.05	0.14	0.18	1.37
150 to 300mm girth	m	0.11	1.65	0.30	0.29	2.24
railings, fences and gates ornamental type					
over 300mm girth	m2	0.26	3.90	0.86	0.71	5.47
not exceeding 150mm girth	m	0.06	0.90	0.14	0.16	1.20
150 to 300mm girth	m	0.10	1.50	0.30	0.27	2.07

One coat green preserver on sawn wood surfaces

	Unit	Labour Hours	Mat'ls £	O & P £	Total £
over 300mm girth	m2	0.38	5.70	0.93	0.99	7.62
not exceeding 300mm girth	m	0.13	1.95	0.30	0.34	2.59
isolated areas not exceeding 0.5m2	nr	0.19	2.85	0.47	0.50	3.82
railings, fences and gates plain open type					
over 300mm girth	m2	0.30	4.50	0.93	0.81	6.24
not exceeding 150mm girth	m	0.07	1.05	0.16	0.18	1.39
150 to 300mm girth	m	0.11	1.65	0.30	0.29	2.24
railings, fences and gates ornamental type					
over 300mm girth	m2	0.26	3.90	0.93	0.72	5.55
not exceeding 150mm girth	m	0.06	0.90	0.16	0.16	1.22
150 to 300mm girth	m	0.10	1.50	0.30	0.27	2.07

	Unit	Labour Hours	Mat'ls £	O & P £	Total £

One coat green preserver
on wrot wood surfaces

	Unit	Labour Hours £	Mat'ls £	O & P £	Total £
over 300mm girth	m2	0.36	5.40	0.86	0.94	7.20

	Unit	Labour	Hours £	Mat'ls £	O & P £	Total £
over 300mm girth	m2	0.36	5.40	0.86	0.94	7.20
not exceeding 300mm girth	m	0.12	1.80	0.25	0.31	2.36
isolated areas not exceeding 0.5m2	nr	0.18	2.70	0.44	0.47	3.61
railings, fences and gates plain open type						
over 300mm girth	m2	0.30	4.50	0.86	0.80	6.16
not exceeding 150mm girth	m	0.07	1.05	0.14	0.18	1.37
150 to 300mm girth	m	0.11	1.65	0.30	0.29	2.24
railings, fences and gates ornamental type						
over 300mm girth	m2	0.26	3.90	0.86	0.71	5.47
not exceeding 150mm girth	m	0.06	0.90	0.14	0.16	1.20
150 to 300mm girth	m	0.10	1.50	0.30	0.27	2.07

FINISHING COATS

One coat white oil-based gloss
finishing coat on undercoated
surfaces over 300mm girth

	Unit	Labour	Hours £	Mat'ls £	O & P £	Total £
general surfaces						
over 300mm girth	m2	0.20	3.00	0.54	0.53	4.07
not exceeding 150mm girth	m	0.06	0.90	0.09	0.15	1.14
150 to 300mm girth	m	0.10	1.50	0.20	0.26	1.96
isolated areas not exceeding 0.5m2	nr	0.12	1.80	0.28	0.31	2.39

	Unit	Labour	Hours £	Mat'ls £	O & P £	Total £
windows, screens glazed doors and the like						
panes area not exceeding 0.2m2	m2	0.37	5.55	0.39	0.89	6.83
panes area 0.1 to 0.5m2	m2	0.32	4.80	0.35	0.77	5.92
panes area 0.5 to 1m2	m2	0.27	4.05	0.32	0.66	5.03
panes area exceeding 1m2	m2	0.24	3.60	0.26	0.58	4.44
frames and linings						
over 300mm girth	m2	0.28	4.20	0.54	0.71	5.45
not exceeding 150mm girth	m	0.06	0.90	0.09	0.15	1.14
150 to 300mm girth	m	0.10	1.50	0.20	0.26	1.96
structural metalwork, general surfaces						
over 300mm girth	m2	0.28	4.20	0.54	0.71	5.45
not exceeding 150mm girth	m	0.06	0.90	0.09	0.15	1.14
150 to 300mm girth	m	0.10	1.50	0.20	0.26	1.96
isolated areas not exceeding 0.5m2	nr	0.13	1.95	0.28	0.33	2.56
windows, screens glazed doors and the like						
panes area not exceeding 0.2m2	m2	0.37	5.55	0.39	0.89	6.83
panes area 0.1 to 0.5m2	m2	0.32	4.80	0.35	0.77	5.92
panes area 0.5 to 1m2	m2	0.27	4.05	0.32	0.66	5.03
panes area exceeding 1m2	m2	0.24	3.60	0.26	0.58	4.44
frames and linings						
over 300mm girth	m2	0.28	4.20	0.54	0.71	5.45
not exceeding 150mm girth	m	0.06	0.90	0.09	0.15	1.14
150 to 300mm girth	m	0.10	1.50	0.20	0.26	1.96

	Unit	Labour	Hours £	Mat'ls £	O & P £	Total £

One coat white oil—based gloss finishing coat (cont'd)

railings, fences and gates
  plain open type

	Unit	Labour	Hours £	Mat'ls £	O & P £	Total £
over 300mm girth	m2	0.28	4.20	0.44	0.70	5.34
not exceeding 150mm girth	m	0.06	0.90	0.08	0.15	1.13
150 to 300mm girth	m	0.11	1.65	0.16	0.27	2.08

railings, fences and gates
  ornamental type

	Unit	Labour	Hours £	Mat'ls £	O & P £	Total £
over 300mm girth	m2	0.26	3.90	0.54	0.67	5.11
not exceeding 150mm girth	m	0.05	0.75	0.14	0.13	1.02
150 to 300mm girth	m	0.10	1.50	0.16	0.25	· 1.91

eaves gutters

	Unit	Labour	Hours £	Mat'ls £	O & P £	Total £
over 300mm girth	m2	0.27	4.05	0.44	0.67	5.16
not exceeding 300mm girth	m	0.10	1.50	0.16	0.25	1.91
isolated areas not exceeding 0.5m2	nr	0.14	2.10	0.23	0.35	2.68

services, pipes, conduits
ducting and the like

	Unit	Labour	Hours £	Mat'ls £	O & P £	Total £
over 300mm girth	m2	0.27	4.05	0.44	0.67	5.16
not exceeding 300mm girth	m	0.10	1.50	0.16	0.25	1.91
isolated areas not exceeding 0.5m2	nr	0.14	2.10	0.23	0.35	2.68

	Unit	Labour Hours	Mat'ls £	O & P £	Total £
One coat coloured oil-based gloss finishing coat on under-coated surfaces over 300mm girth					
general surfaces					
over 300mm girth	m2	0.20	3.00	0.56	0.53	4.09
not exceeding 150mm girth	m	0.06	0.90	0.10	0.15	1.15
150 to 300mm girth	m	0.10	1.50	0.21	0.26	1.97
isolated areas not exceeding 0.5m2	nr	0.12	1.80	0.30	0.32	2.42
windows, screens glazed doors and the like					
panes area not exceeding 0.2m2	m2	0.37	5.55	0.41	0.89	6.85
panes area 0.1 to 0.5m2	m2	0.32	4.80	0.37	0.78	5.95
panes area 0.5 to 1m2	m2	0.27	4.05	0.33	0.66	5.04
panes area exceeding 1m2	m2	0.24	3.60	0.29	0.58	4.47
frames and linings					
over 300mm girth	m2	0.28	4.20	0.56	0.71	5.47
not exceeding 150mm girth	m	0.06	0.90	0.10	0.15	1.15
150 to 300mm girth	m	0.10	1.50	0.21	0.26	1.97
structural metalwork, general surfaces					
over 300mm girth	m2	0.37	5.55	0.56	0.92	7.03
not exceeding 150mm girth	m	0.06	0.90	0.10	0.15	1.15
150 to 300mm girth	m	0.24	3.60	0.21	0.57	4.38
isolated areas not exceeding 0.5m2	nr	0.12	1.80	0.29	0.31	2.40

	Unit	Labour	Hours £	Mat'ls £	O & P £	Total £
One coat coloured oil-based gloss finishing coat (cont'd)						
eaves gutters						
over 150mm girth	m2	0.27	4.05	0.47	0.68	5.20
not exceeding 300mm girth	m	0.50	7.50	0.18	1.15	8.83
isolated areas not exceeding 0.5m2	nr	0.14	2.10	0.25	0.35	2.70
services, pipes, conduits ducting and the like						
over 300mm girth	m2	0.27	4.05	0.47	0.68	5.20
not exceeding 300mm girth	m	0.10	1.50	0.18	0.25	1.93
isolated areas not exceeding 0.5m2	nr	0.14	2.10	0.25	0.35	2.70
railings, fences and gates plain open type						
over 300mm girth	m2	0.28	4.20	0.56	0.71	5.47
not exceeding 150mm girth	m	0.06	0.90	0.08	0.15	1.13
150 to 300mm girth	m	0.11	1.65	0.16	0.27	2.08
railings, fences and gates ornamental type						
over 300mm girth	m2	0.26	3.90	0.56	0.67	5.13
not exceeding 150mm girth	m	0.05	0.75	0.08	0.12	0.95
150 to 300mm girth	m	0.10	1.50	0.16	0.25	1.91

	Unit	Labour Hours	Mat'ls £	O & P £	Total £
One coat Snowcem on primed surfaces over 300mm girth					
brickwork walls	m2	0.12	1.80	0.68	0.37	2.85
blockwork walls	m2	0.14	2.10	0.68	0.42	3.20
concrete walls	m2	0.11	1.65	0.64	0.34	2.63
rendered walls	m2	0.12	1.80	0.64	0.37	2.81
roughcast walls	m2	0.14	2.10	0.77	0.43	3.30
One coat Sandtex matt on primed surfaces over 300mm girth					
brickwork walls	m2	0.12	1.80	1.32	0.47	3.59
blockwork walls	m2	0.14	2.10	1.32	0.51	3.93
concrete walls	m2	0.11	1.65	1.23	0.43	3.31
rendered walls	m2	0.12	1.80	1.23	0.45	3.48
roughcast walls	m2	0.14	2.10	1.43	0.53	4.06
One coat Sandtex textured finish on surfaces over 300mm girth					
brickwork walls	m2	0.12	1.80	3.76	0.83	6.39
blockwork walls	m2	0.14	2.10	3.76	0.88	6.74
concrete walls	m2	0.11	1.65	3.56	0.78	5.99
rendered walls	m2	0.12	1.80	3.56	0.80	6.16
roughcast walls	m2	0.14	2.10	3.98	0.91	6.99
One coat clear preserver on sawn wood surfaces					
over 300mm girth	m2	0.38	5.70	0.94	1.00	7.64
not exceeding 300mm girth	m	0.13	1.95	0.30	0.34	2.59
isolated areas not exceeding 0.5m2	nr	0.19	2.85	0.48	0.50	3.83

	Unit	Labour	Hours £	Mat'ls £	O & P £	Total £
One coat clear preserver (cont'd)						
railings, fences and gates plain open type						
over 300mm girth	m2	0.30	4.50	0.96	0.82	6.28
not exceeding 150mm girth	m	0.07	1.05	0.16	0.18	1.39
150 to 300mm girth	m	0.11	1.65	0.30	0.29	2.24
railings, fences and gates ornamental type						
over 300mm girth	m2	0.26	3.90	0.96	0.73	5.59
not exceeding 150mm girth	m	0.06	0.90	0.16	0.16	1.22
150 to 300mm girth	m	0.10	1.50	0.30	0.27	2.07
One coat clear preserver on wrot wood surfaces						
over 300mm girth	m2	0.38	5.70	0.96	1.00	7.66
not exceeding 300mm girth	m	0.60	9.00	0.30	1.40	10.70
isolated areas not exceeding 0.5m2	nr	0.19	2.85	0.47	0.50	3.82
railings, fences and gates ornamental type						
over 300mm girth	m2	0.30	4.50	0.96	0.82	6.28
not exceeding 150mm girth	m	0.07	1.05	0.16	0.18	1.39
150 to 300mm girth	m	0.11	1.65	0.30	0.29	2.24

	Unit	Labour	Hours £	Mat'ls £	O & P £	Total £
railings, fences and gates ornamental type						
over 300mm girth	m2	0.26	3.90	0.96	0.73	5.59
not exceeding 150mm girth	m	0.06	0.90	0.16	0.16	1.22
150 to 300mm girth	m	0.10	1.50	0.30	0.27	2.07
One coat light preserver on sawn wood surfaces						
over 300mm girth	m2	0.38	5.70	0.96	1.00	7.66
not exceeding 300mm girth	m	0.13	1.95	0.30	0.34	2.59
isolated areas not exceeding 0.5m2	nr	0.19	2.85	0.47	0.50	3.82
railings, fences and gates plain open type						
over 300mm girth	m2	0.30	4.50	0.96	0.82	6.28
not exceeding 150mm girth	m	0.07	1.05	0.16	0.18	1.39
150 to 300mm girth	m	0.11	1.65	0.30	0.29	2.24
railings, fences and gates ornamental type						
over 300mm girth	m2	0.26	3.90	0.96	0.73	5.59
not exceeding 150mm girth	m	0.06	0.90	0.16	0.16	1.22
150 to 300mm girth	m	0.10	1.50	0.30	0.27	2.07
One coat light preserver on wrot wood surfaces						
over 300mm girth	m2	0.36	5.40	0.88	0.94	7.22
not exceeding 300mm girth	m	0.12	1.80	0.27	0.31	2.38

	Unit	Labour Hours	Mat'ls £	O & P £	Total £
		£			

One coat light preserver (cont'd)

isolated areas not exceeding 0.5m2	nr	0.18	2.70	0.45	0.47	3.62

railings, fences and gates
  plain open type

over 300mm girth	m2	0.30	4.50	0.96	0.82	6.28
not exceeding 150mm girth	m	0.07	1.05	0.16	0.18	1.39
150 to 300mm girth	m	0.11	1.65	0.30	0.29	2.24

railings, fences and gates
  ornamental type

over 300mm girth	m2	0.26	3.90	0.96	0.73	5.59
not exceeding 150mm girth	m	0.06	0.90	0.16	0.16	1.22
150 to 300mm girth	m	0.10	1.50	0.30	0.27	2.07

One coat dark preserver on sawn wood surfaces

over 300mm girth	m2	0.38	5.70	0.96	1.00	7.66
not exceeding 300mm girth	m	0.13	1.95	0.30	0.34	2.59
isolated areas not exceeding 0.5m2	nr	0.19	2.85	0.47	0.50	3.82

railings, fences and gates
  plain open type

over 300mm girth	m2	0.30	4.50	0.96	0.82	6.28
not exceeding 150mm girth	m	0.07	1.05	0.16	0.18	1.39
150 to 300mm girth	m	0.11	1.65	0.30	0.29	2.24

	Unit	Labour Hours	Mat'ls £	O & P £	Total £
railings, fences and gates ornamental type					
over 300mm girth	m2	0.26 3.90	0.96	0.73	5.59
not exceeding 150mm girth	m	0.06 0.90	0.16	0.16	1.22
150 to 300mm girth	m	0.10 1.50	0.30	0.27	2.07
One coat dark preserver on wrot wood surfaces					
over 300mm girth	m2	0.36 5.40	0.88	0.94	7.22
not exceeding 300mm girth	m	0.12 1.80	0.27	0.31	2.38
isolated areas not exceeding 0.5m2	nr	0.18 2.70	0.45	0.47	3.62
railings, fences and gates plain open type					
over 300mm girth	m2	0.30 4.50	0.96	0.82	6.28
not exceeding 150mm girth	m	0.07 1.05	0.16	0.18	1.39
150 to 300mm girth	m	0.11 1.65	0.30	0.29	2.24
railings, fences and gates ornamental type					
over 300mm girth	m2	0.26 3.90	0.96	0.73	5.59
not exceeding 150mm girth	m	0.06 0.90	0.16	0.16	1.22
150 to 300mm girth	m	0.10 1.50	0.30	0.27	2.07
One coat green preserver on sawn wood surfaces					
over 300mm girth	m2	0.38 5.70	0.96	1.00	7.66
not exceeding 150mm girth	m	0.06 0.90	0.16	0.16	1.22

	Unit	Labour Hours	Mat'ls £	O & P £	Total £
			£	£	£

One coat green preserver (cont'd)

150 to 300mm girth	m	0.10	1.50	0.30	0.27	2.07

railings, fences and gates
plain open type

over 300mm girth	m2	0.30	4.50	0.96	0.82	6.28
not exceeding 150mm girth	m	0.07	1.05	0.16	0.18	1.39
150 to 300mm girth	m	0.11	1.65	0.30	0.29	2.24

railings, fences and gates
ornamental type

over 300mm girth	m2	0.26	3.90	0.96	0.73	5.59
not exceeding 150mm girth	m	0.06	0.90	0.16	0.16	1.22
150 to 300mm girth	m	0.10	1.50	0.30	0.27	2.07

One coat green preserver on wrot wood surfaces

over 300mm girth	m2	0.36	5.40	0.88	0.94	7.22
not exceeding 300mm girth	m	0.12	1.80	0.27	0.31	2.38
isolated areas not exceeding 0.5m2	nr	0.18	2.70	0.45	0.47	3.62

railings, fences and gates
plain open type

over 300mm girth	m2	0.30	4.50	0.96	0.82	6.28
not exceeding 150mm girth	m	0.07	1.05	0.16	0.18	1.39
150 to 300mm girth	m	0.11	1.65	0.30	0.29	2.24

	Unit	Labour	Hours £	Mat'ls £	O & P £	Total £
railings, fences and gates ornamental type						
over 300mm girth	m2	0.26	3.90	0.96	0.73	5.59
not exceeding 150mm girth	m	0.06	0.90	0.16	0.16	1.22
150 to 300mm girth	m	0.10	1.50	0.30	0.27	2.07

Part Four

COMPOSITE RATES – PAINTING AND WALLPAPERING

Internally

Externally

	Unit	Labour Hours £	Mat'ls £	O & P £	Total £

The following rates are based upon the rates in Part 3 – Unit Rates for Painting and Wall-papering and apply where more than one painting operation is required.

INTERNALLY

One mist coat and two coats white matt emulsion on surfaces over 300mm girth

	Unit	Labour Hours £	Mat'ls £	O & P £	Total £
brickwork walls	m2	0.26	3.90	1.45	0.80	6.15
brickwork walls in staircase areas	m2	0.32	4.80	1.45	0.94	7.19
blockwork walls	m2	0.32	4.80	1.65	0.97	7.42
blockwork walls in staircase areas	m2	0.38	5.70	1.65	1.10	8.45
concrete walls	m2	0.23	3.45	1.30	0.71	5.46
concrete walls in staircase areas	m2	0.29	4.35	1.30	0.85	6.50
concrete ceilings	m2	0.35	5.25	1.30	0.98	7.53
concrete ceilings in staircase areas	m2	0.21	3.15	1.30	0.67	5.12
plastered walls	m2	0.20	3.00	1.30	0.65	4.95
plastered walls in staircase areas	m2	0.26	3.90	1.30	0.78	5.98
plastered ceilings	m2	0.26	3.90	1.30	0.78	5.98
plastered ceilings in staircase areas	m2	0.32	4.80	1.30	0.92	7.02
embossed paper	m2	0.20	3.00	1.30	0.65	4.95
embossed paper in staircase areas	m2	0.22	3.30	1.30	0.69	5.29

	Unit	Labour	Hours £	Mat'ls £	O & P £	Total £
One mist coat and two coats coloured matt emulsion on surfaces over 300mm girth						
brickwork walls	m2	0.26	3.90	1.55	0.82	6.27
brickwork walls in staircase areas	m2	0.32	4.80	1.55	0.95	7.30
blockwork walls	m2	0.32	4.80	1.75	0.98	7.53
blockwork walls in staircase areas	m2	0.38	5.70	1.75	1.12	8.57
concrete walls	m2	0.23	3.45	1.40	0.73	5.58
concrete walls in staircase areas	m2	0.29	4.35	1.41	0.86	6.62
concrete ceilings	m2	0.35	5.25	1.41	1.00	7.66
concrete ceilings in staircase areas	m2	0.21	3.15	1.41	0.68	5.24
plastered walls	m2	0.20	3.00	1.41	0.66	5.07
plastered walls in staircase areas	m2	0.26	3.90	1.41	0.80	6.11
plastered ceilings	m2	0.26	3.90	1.41	0.80	6.11
plastered ceilings in staircase areas	m2	0.32	4.80	1.41	0.93	7.14
embossed paper	m2	0.20	3.00	1.41	0.66	5.07
embossed paper in staircase areas	m2	0.26	3.90	1.41	0.80	6.11
One mist coat and two coats white silk emulsion on surfaces over 300mm girth						
brickwork walls	m2	0.26	3.90	1.55	0.82	6.27
brickwork walls in staircase areas	m2	0.32	4.80	1.55	0.95	7.30
blockwork walls	m2	0.32	4.80	1.75	0.98	7.53

	Unit	Labour	Hours £	Mat'ls £	O & P £	Total £
blockwork walls in staircase areas	m2	0.28	4.20	1.45	0.85	6.50
concrete walls	m2	0.23	3.45	1.45	0.74	5.64
concrete walls in staircase areas	m2	0.29	4.35	1.45	0.87	6.67
concrete ceilings	m2	0.35	5.25	1.45	1.01	7.71
concrete ceilings in staircase areas	m2	0.21	3.15	1.45	0.69	5.29
plastered walls	m2	0.20	3.00	1.45	0.67	5.12
plastered walls in staircase areas	m2	0.26	3.90	1.45	0.80	6.15
plastered ceilings	m2	0.26	3.90	1.45	0.80	6.15
plastered ceilings in staircase areas	m2	0.32	4.80	1.45	0.94	7.19
embossed paper	m2	0.20	3.00	1.45	0.67	5.12
embossed paper in staircase areas	m2	0.26	3.90	1.45	0.80	6.15

One mist coat and two coats coloured silk emulsion on surfaces over 300mm girth

	Unit	Labour	Hours £	Mat'ls £	O & P £	Total £
brickwork walls	m2	0.26	3.90	1.70	0.84	6.44
brickwork walls in staircase areas	m2	0.32	4.80	1.70	0.98	7.48
blockwork walls	m2	0.32	4.80	1.90	1.01	7.71
blockwork walls in staircase areas	m2	0.28	4.20	1.55	0.86	6.61
concrete walls	m2	0.23	3.45	1.55	0.75	5.75
concrete walls in staircase areas	m2	0.29	4.35	1.55	0.89	6.79
concrete ceilings	m2	0.35	5.25	1.55	1.02	7.82
concrete ceilings in staircase areas	m2	0.21	3.15	1.55	0.71	5.41

	Unit	Labour	Hours £	Mat'ls £	O & P £	Total £
One mist coat and two coats coloured silk emulsion (cont'd)						
plastered walls	m2	0.20	3.00	1.55	0.68	5.23
plastered walls in staircase areas	m2	0.26	3.90	1.55	0.82	6.27
plastered ceilings	m2	0.26	3.90	1.55	0.82	6.27
plastered ceilings in staircase areas	m2	0.32	4.80	1.55	0.95	7.30
embossed paper	m2	0.20	3.00	1.55	0.68	5.23
embossed paper in staircase areas	m2	0.26	3.90	1.55	0.82	6.27
One coat alkali-resisting primer and two coats white matt emulsion on surfaces over 300mm girth						
brickwork walls	m2	0.26	3.90	2.35	0.94	7.19
brickwork walls in staircase areas	m2	0.32	4.80	2.35	1.07	8.22
blockwork walls	m2	0.32	4.80	3.00	1.17	8.97
blockwork walls in staircase areas	m2	0.28	4.20	3.00	1.08	8.28
concrete walls	m2	0.23	3.45	2.02	0.82	6.29
concrete walls in staircase areas	m2	0.29	4.35	2.02	0.96	7.33
concrete ceilings	m2	0.35	5.25	2.02	1.09	8.36
concrete ceilings in staircase areas	m2	0.21	3.15	2.02	0.78	5.95
plastered walls	m2	0.20	3.00	2.02	0.75	5.77
plastered walls in staircase areas	m2	0.26	3.90	2.02	0.89	6.81

	Unit	Labour	Hours £	Mat'ls £	O & P £	Total £
plastered ceilings	m2	0.26	3.90	2.02	0.89	6.81
plastered ceilings in staircase areas	m2	0.32	4.80	2.02	1.02	7.84
embossed paper	m2	0.20	3.00	2.02	0.75	5.77
embossed paper in staircase areas	m2	0.26	3.90	2.02	0.89	6.81

One coat alkali-resisting primer and two coats coloured matt emulsion on surfaces over 300mm girth

	Unit	Labour	Hours £	Mat'ls £	O & P £	Total £
brickwork walls	m2	0.26	3.90	2.44	0.95	7.29
brickwork walls in staircase areas	m2	0.32	4.80	2.44	1.09	8.33
blockwork walls	m2	0.32	4.80	3.05	1.18	9.03
blockwork walls in staircase areas	m2	0.28	4.20	3.05	1.09	8.34
concrete walls	m2	0.23	3.45	2.10	0.83	6.38
concrete walls in staircase areas	m2	0.29	4.35	2.10	0.97	7.42
concrete ceilings	m2	0.35	5.25	2.10	1.10	8.45
concrete ceilings in staircase areas	m2	0.21	3.15	2.10	0.79	6.04
plastered walls	m2	0.20	3.00	2.10	0.77	5.87
plastered walls in staircase areas	m2	0.26	3.90	2.10	0.90	6.90
plastered ceilings	m2	0.26	3.90	2.10	0.90	6.90
plastered ceilings in staircase areas	m2	0.32	4.80	2.10	1.04	7.94
embossed paper	m2	0.20	3.00	2.10	0.77	5.87
embossed paper in staircase areas	m2	0.26	3.90	2.10	0.90	6.90

	Unit	Labour	Hours £	Mat'ls £	O & P £	Total £
One coat alkali-resisting primer and two coats white silk emulsion on surfaces over 300mm girth						
brickwork walls	m2	0.26	3.90	2.12	0.90	6.92
brickwork walls in staircase areas	m2	0.32	4.80	2.12	1.04	7.96
blockwork walls	m2	0.32	4.80	3.10	1.19	9.09
blockwork walls in staircase areas	m2	0.28	4.20	3.10	1.10	8.40
concrete walls	m2	0.23	3.45	2.12	0.84	6.41
concrete walls in staircase areas	m2	0.29	4.35	2.12	0.97	7.44
concrete ceilings	m2	0.35	5.25	2.12	1.11	8.48
concrete ceilings in staircase areas	m2	0.21	3.15	2.12	0.79	6.06
plastered walls	m2	0.20	3.00	2.12	0.77	5.89
plastered walls in staircase areas	m2	0.26	3.90	2.12	0.90	6.92
plastered ceilings	m2	0.26	3.90	2.12	0.90	6.92
plastered ceilings in staircase areas	m2	0.32	4.80	2.12	1.04	7.96
embossed paper	m2	0.20	3.00	2.12	0.77	5.89
embossed paper in staircase areas	m2	0.26	3.90	2.12	0.90	6.92
One coat alkali-resisting primer and two coats coloured silk emulsion on surfaces over 300mm girth						
brickwork walls	m2	0.26	3.90	2.60	0.98	7.48
brickwork walls in staircase areas	m2	0.32	4.80	2.60	1.11	8.51
blockwork walls	m2	0.32	4.80	3.20	1.20	9.20

	Unit	Labour Hours £	Mat'ls £	O & P £	Total £
blockwork walls in staircase areas	m2	0.28	4.20	3.20	1.11	8.51
concrete walls	m2	0.23	3.45	2.25	0.86	6.56
concrete walls in staircase areas	m2	0.29	4.35	2.25	0.99	7.59
concrete ceilings	m2	0.35	5.25	2.25	1.13	8.63
concrete ceilings in staircase areas	m2	0.21	3.15	2.25	0.81	6.21
plastered walls	m2	0.20	3.00	2.25	0.79	6.04
plastered walls in staircase areas	m2	0.26	3.90	2.25	0.92	7.07
plastered ceilings	m2	0.26	3.90	2.25	0.92	7.07
plastered ceilings in staircase areas	m2	0.32	4.80	2.25	1.06	8.11
embossed paper	m2	0.20	3.00	2.25	0.79	6.04
embossed paper in staircase areas	m2	0.26	3.90	2.25	0.92	7.07

One coat wood primer, one white oil-based undercoat and one white oil-based finishing coat on wood

general surfaces					
over 300mm girth	m2	0.50	7.50	1.76	1.39	10.65
not exceeding 150mm girth	m	0.08	1.20	0.30	0.23	1.73
150 to 300mm girth	m	0.18	2.70	0.65	0.50	3.85
isolated areas not exceeding 0.5m2	nr	0.34	5.10	0.88	0.90	6.88

windows, screens glazed doors and the like

panes area not exceeding 0.2m2	m2	0.25	3.75	1.10	0.73	5.58

	Unit	Labour Hours	Mat'ls £	O & P £	Total £

One coat wood primer, one white oil-based undercoat and one white oil-based finishing coat (cont'd)

	Unit	Labour Hours	Mat'ls £	O & P £	Total £
panes area 0.1 to 0.5m2	m2	0.86	12.90	0.96	2.08	15.94
panes area 0.5 to 1m2	m2	0.71	10.65	0.85	1.73	13.23
panes area exceeding 1m2	m2	0.62	9.30	0.75	1.51	11.56
frames and linings						
over 300mm girth	m2	0.54	8.10	1.76	1.48	11.34
not exceeding 150mm girth	m	0.10	1.50	0.30	0.27	2.07
150 to 300mm girth	m	0.20	3.00	0.65	0.55	4.20
skirtings and rails						
over 300mm girth	m2	0.54	8.10	1.76	1.48	11.34
not exceeding 150mm girth	m	0.10	1.50	0.30	0.27	2.07
150 to 300mm girth	m	0.20	3.00	0.65	0.55	4.20

One coat wood primer, one coloured oil-based undercoat and one coloured oil-based finishing coat on wood

	Unit	Labour Hours	Mat'ls £	O & P £	Total £
general surfaces						
over 300mm girth	m2	0.50	7.50	1.72	1.38	10.60
not exceeding 150mm girth	m	0.08	1.20	0.30	0.23	1.73
150 to 300mm girth	m	0.18	2.70	0.60	0.50	3.80
isolated areas not exceeding 0.5m2	nr	0.25	3.75	0.88	0.69	5.32

	Unit	Labour Hours	£	Mat'ls £	O & P £	Total £
windows, screens glazed doors and the like						
panes area not exceeding 0.2m2	m2	1.01	15.15	1.08	2.43	18.66
panes area 0.1 to 0.5m2	m2	0.86	12.90	0.95	2.08	15.93
panes area 0.5 to 1m2	m2	0.71	10.65	0.80	1.72	13.17
panes area exceeding 1m2	m2	0.62	9.30	0.70	1.50	11.50
frames and linings						
over 300mm girth	m2	0.54	8.10	1.72	1.47	11.29
not exceeding 150mm girth	m	0.10	1.50	0.30	0.27	2.07
150 to 300mm girth	m	0.20	3.00	0.60	0.54	4.14
skirtings and rails						
over 300mm girth	m2	0.54	8.10	1.72	1.47	11.29
not exceeding 150mm girth	m	0.10	1.50	0.30	0.27	2.07
150 to 300mm girth	m	0.20	3.00	0.60	0.54	4.14
One coat wood primer, one white oil-based undercoat and one white eggshell finishing coat on wood						
general surfaces						
over 300mm girth	m2	0.50	7.50	1.72	1.38	10.60
not exceeding 150mm girth	m	0.08	1.20	0.30	0.23	1.73
150 to 300mm girth	m	0.18	2.70	0.60	0.50	3.80
isolated areas not exceeding 0.5m2	nr	0.25	3.75	0.88	0.69	5.32

	Unit	Labour Hours	Mat'ls £	O & P £	Total £

One coat wood primer, one white oil-based undercoat and one white eggshell finishing coat (cont'd)

windows, screens glazed doors and the like
panes area not exceeding

	Unit	Labour Hours	Mat'ls £	O & P £	Total £
0.2m2	m2	1.01	15.15	1.08	2.43	18.66
panes area 0.1 to 0.5m2	m2	0.86	12.90	0.95	2.08	15.93
panes area 0.5 to 1m2	m2	0.71	10.65	0.80	1.72	13.17
panes area exceeding 1m2	m2	0.62	9.30	0.70	1.50	11.50
frames and linings						
over 300mm girth	m2	0.54	8.10	1.72	1.47	11.29
not exceeding 150mm girth	m	0.10	1.50	0.30	0.27	2.07
150 to 300mm girth	m	0.20	3.00	0.60	0.54	4.14
skirtings and rails						
over 300mm girth	m2	0.54	8.10	1.72	1.47	11.29
not exceeding 150mm girth	m	0.10	1.50	0.30	0.27	2.07
150 to 300mm girth	m	0.20	3.00	0.60	0.54	4.14

One coat wood primer, one coloured oil-based undercoat and one coloured eggshell finishing coat on wood

	Unit	Labour Hours	Mat'ls £	O & P £	Total £
general surfaces						
over 300mm girth	m2	0.50	7.50	1.78	1.39	10.67
not exceeding 150mm girth	m	0.08	1.20	0.30	0.23	1.73
150 to 300mm girth	m	0.18	2.70	0.62	0.50	3.82

	Unit	Labour	Hours £	Mat'ls £	O & P £	Total £
isolated areas not exceeding 0.5m2	nr	0.25	3.75	0.90	0.70	5.35
windows, screens glazed doors and the like						
panes area not exceeding 0.2m2	m2	1.01	15.15	1.14	2.44	18.73
panes area 0.1 to 0.5m2	m2	0.86	12.90	1.00	2.09	15.99
panes area 0.5 to 1m2	m2	0.71	10.65	0.88	1.73	13.26
panes area exceeding 1m2	m2	0.62	9.30	0.74	1.51	11.55
frames and linings						
over 300mm girth	m2	0.54	8.10	1.78	1.48	11.36
not exceeding 150mm girth	m	0.10	1.50	0.30	0.27	2.07
150 to 300mm girth	m	0.20	3.00	0.62	0.54	4.16
skirtings and rails						
over 300mm girth	m2	0.54	8.10	1.78	1.48	11.36
not exceeding 150mm girth	m	0.10	1.50	0.30	0.27	2.07
150 to 300mm girth	m	0.20	3.00	0.62	0.54	4.16

One coat aluminium primer, one white oil-based undercoat and one white oil-based finishing coat on wood

	Unit	Labour	Hours £	Mat'ls £	O & P £	Total £
general surfaces						
over 300mm girth	m2	0.50	7.50	2.40	1.49	11.39
not exceeding 150mm girth	m	0.08	1.20	0.40	0.24	1.84
150 to 300mm girth	m	0.18	2.70	0.82	0.53	4.05
isolated areas not exceeding 0.5m2	nr	0.25	3.75	1.28	0.75	5.78

	Unit	Labour Hours	Mat'ls £	O & P £	Total £

One coat aluminium primer, one white oil-based undercoat and one white oil-based finishing coat (cont'd)

windows, screens glazed
doors and the like
panes area not exceeding

	Unit	Labour Hours	Mat'ls £	O & P £	Total £
0.2m2	m2	1.01	15.15	1.40	2.48	19.03
panes area 0.1 to 0.5m2	m2	0.86	12.90	1.28	2.13	16.31
panes area 0.5 to 1m2	m2	0.71	10.65	1.02	1.75	13.42
panes area exceeding 1m2	m2	0.62	9.30	0.60	1.49	11.39
frames and linings						
over 300mm girth	m2	0.54	8.10	2.40	1.58	12.08
not exceeding 150mm girth	m	0.10	1.50	0.40	0.29	2.19
150 to 300mm girth	m	0.20	3.00	0.82	0.57	4.39
skirtings and rails						
over 300mm girth	m2	0.54	8.10	2.40	1.58	12.08
not exceeding 150mm girth	m	0.10	1.50	0.40	0.29	2.19
150 to 300mm girth	m	0.20	3.00	0.82	0.57	4.39

One coat aluminium primer, one coloured oil-based undercoat and one coloured oil-based finishing coat on wood

general surfaces	Unit	Labour Hours	Mat'ls £	O & P £	Total £
over 300mm girth	m2	0.50	7.50	2.44	1.49	11.43
not exceeding 150mm girth	m	0.08	1.20	0.42	0.24	1.86

	Unit	Labour Hours	Mat'ls £	O & P £	Total £
150 to 300mm girth	m	0.18	2.70	0.84	0.53	4.07
isolated areas not exceeding 0.5m2	nr	0.25	3.75	1.24	0.75	5.74

windows, screens glazed doors and the like

panes area not exceeding 0.2m2	m2	1.01	15.15	1.44	2.49	19.08
panes area 0.1 to 0.5m2	m2	0.86	12.90	1.30	2.13	16.33
panes area 0.5 to 1m2	m2	0.71	10.65	1.16	1.77	13.58
panes area exceeding 1m2	m2	0.62	9.30	1.04	1.55	11.89

frames and linings

over 300mm girth	m2	0.54	8.10	2.44	1.58	12.12
not exceeding 150mm girth	m	0.10	1.50	0.42	0.29	2.21
150 to 300mm girth	m	0.20	3.00	0.78	0.57	4.35

skirtings and rails

over 300mm girth	m2	0.54	8.10	2.44	1.58	12.12
not exceeding 150mm girth	m	0.10	1.50	0.42	0.29	2.21
150 to 300mm girth	m	0.20	3.00	0.78	0.57	4.35

One coat aluminium primer, one white oil-based undercoat and one white eggshell finishing coat on wood

general surfaces

over 300mm girth	m2	0.50	7.50	2.32	1.47	11.29
not exceeding 150mm girth	m	0.08	1.20	0.40	0.24	1.84
150 to 300mm girth	m	0.18	2.70	0.78	0.52	4.00
isolated areas not exceeding 0.5m2	nr	0.25	3.75	1.18	0.74	5.67

	Unit	Labour Hours	Mat'ls £	O & P £	Total £

**One coat aluminium primer,
one white oil-based undercoat
and one white egshell
finishing coat (cont'd)**

windows, screens glazed
doors and the like
panes area not exceeding

	Unit	Labour Hours	Mat'ls £	O & P £	Total £
0.2m2	m2	1.01	15.15	1.36	2.48	18.99
panes area 0.1 to 0.5m2	m2	0.86	12.90	1.22	2.12	16.24
panes area 0.5 to 1m2	m2	0.71	10.65	1.10	1.76	13.51
panes area exceeding 1m2	m2	0.62	9.30	0.96	1.54	11.80
frames and linings						
over 300mm girth	m2	0.54	8.10	2.32	1.56	11.98
not exceeding 150mm girth	m	0.10	1.50	0.40	0.29	2.19
150 to 300mm girth	m	0.20	3.00	0.78	0.57	4.35
skirtings and rails						
over 300mm girth	m2	0.54	8.10	2.32	1.56	11.98
not exceeding 150mm girth	m	0.10	1.50	0.40	0.29	2.19
150 to 300mm girth	m	0.20	3.00	0.78	0.57	4.35

**One coat aluminium primer,
one coloured oil-based undercoat
and one coloured eggshell
finishing coat on wood**

	Unit	Labour Hours	Mat'ls £	O & P £	Total £
general surfaces						
over 300mm girth	m2	0.50	7.50	2.40	1.49	11.39
not exceeding 150mm girth	m	0.08	1.20	0.50	0.26	1.96

	Unit	Labour	Hours £	Mat'ls £	O & P £	Total £
150 to 300mm girth isolated areas not	m	0.18	2.70	0.94	0.55	4.19
exceeding 0.5m2	nr	0.25	3.75	1.30	0.76	5.81
windows, screens glazed doors and the like panes area not exceeding						
0.2m2	m2	1.01	15.15	1.42	2.49	19.06
panes area 0.1 to 0.5m2	m2	0.86	12.90	1.30	2.13	16.33
panes area 0.5 to 1m2	m2	0.71	10.65	1.16	1.77	13.58
panes area exceeding 1m2	m2	0.62	9.30	1.04	1.55	11.89
frames and linings over 300mm girth	m2	0.54	8.10	2.50	1.59	12.19
not exceeding 150mm girth	m	0.10	1.50	0.42	0.29	2.21
150 to 300mm girth	m	0.20	3.00	0.80	0.57	4.37
skirtings and rails over 300mm girth	m2	0.54	8.10	2.50	1.59	12.19
not exceeding 150mm girth	m	0.10	1.50	0.42	0.29	2.21
150 to 300mm girth	m	0.20	3.00	0.80	0.57	4.37

One coat acrylic primer, one white oil-based undercoat and one white oil-based finishing coat on wood

	Unit	Labour	Hours £	Mat'ls £	O & P £	Total £
general surfaces over 300mm girth	m2	0.50	7.50	1.76	1.39	10.65
not exceeding 150mm girth	m	0.08	1.20	0.40	0.24	1.84
150 to 300mm girth	m	0.18	2.70	0.66	0.50	3.86
isolated areas not exceeding 0.5m2	nr	0.25	3.75	0.88	0.69	5.32

	Unit	Labour Hours	Mat'ls £	O & P £	Total £

One coat acrylic primer,
one white oil-based undercoat
and one white oil-based
finishing coat (cont'd)

windows, screens glazed
doors and the like

	Unit	Labour Hours	Mat'ls £	O & P £	Total £
panes area not exceeding 0.2m2	m2	1.01	15.15	1.14	2.44	18.73
panes area 0.1 to 0.5m2	m2	0.86	12.90	0.96	2.08	15.94
panes area 0.5 to 1m2	m2	0.71	10.65	0.84	1.72	13.21
panes area exceeding 1m2	m2	0.62	9.30	0.72	1.50	11.52
frames and linings						
over 300mm girth	m2	0.54	8.10	1.76	1.48	11.34
not exceeding 150mm girth	m	0.10	1.50	0.40	0.29	2.19
150 to 300mm girth	m	0.20	3.00	0.66	0.55	4.21
skirtings and rails						
over 300mm girth	m2	0.54	8.10	1.76	1.48	11.34
not exceeding 150mm girth	m	0.10	1.50	0.40	0.29	2.19
150 to 300mm girth	m	0.20	3.00	0.66	0.55	4.21

One coat acrylic primer,
one coloured oil-based undercoat
and one coloured oil-based
finishing coat on wood

general surfaces

	Unit	Labour Hours	Mat'ls £	O & P £	Total £
over 300mm girth	m2	0.50	7.50	1.76	1.39	10.65
not exceeding 150mm girth	m	0.08	1.20	0.40	0.24	1.84

	Unit	Labour Hours	£	Mat'ls £	O & P £	Total £
150 to 300mm girth	m	0.18	2.70	0.66	0.50	3.86
isolated areas not						
exceeding 0.5m2	nr	0.25	3.75	0.90	0.70	5.35
windows, screens glazed						
doors and the like						
panes area not exceeding						
0.2m2	m2	1.01	15.15	1.12	2.44	18.71
panes area 0.1 to 0.5m2	m2	0.86	12.90	1.00	2.09	15.99
panes area 0.5 to 1m2	m2	0.71	10.65	0.86	1.73	13.24
panes area exceeding 1m2	m2	0.62	9.30	0.74	1.51	11.55
frames and linings						
over 300mm girth	m2	0.54	8.10	1.80	1.49	11.39
not exceeding 150mm						
girth	m	0.10	1.50	0.30	0.27	2.07
150 to 300mm girth	m	0.20	3.00	0.66	0.55	4.21
skirtings and rails						
over 300mm girth	m2	0.54	8.10	1.80	1.49	11.39
not exceeding 150mm						
girth	m	0.10	1.50	0.30	0.27	2.07
150 to 300mm girth	m	0.20	3.00	0.66	0.55	4.21

One coat acrylic primer,
one white oil-based undercoat
and one white eggshell
finishing coat on wood

	Unit	Labour Hours	£	Mat'ls £	O & P £	Total £
general surfaces						
over 300mm girth	m2	0.50	7.50	1.70	1.38	10.58
not exceeding 150mm						
girth	m	0.16	2.40	0.30	0.41	3.11
150 to 300mm girth	m	0.28	4.20	0.62	0.72	5.54
isolated areas not						
exceeding 0.5m2	nr	0.34	5.10	0.86	0.89	6.85

	Unit	Labour Hours	Mat'ls £	O & P £	Total £

One coat acrylic primer,
one white oil-based undercoat
and one white eggshell
finishing coat (cont'd)

windows, screens glazed
doors and the like
panes area not exceeding

	Unit	Labour Hours	Mat'ls £	O & P £	Total £
0.2m2	m2	1.01	15.15	1.06	2.43	18.64
panes area 0.1 to 0.5m2	m2	0.86	12.90	0.94	2.08	15.92
panes area 0.5 to 1m2	m2	0.71	10.65	0.80	1.72	13.17
panes area exceeding 1m2	m2	0.62	9.30	0.68	1.50	11.48

frames and linings

over 300mm girth	m2	0.54	8.10	1.70	1.47	11.27
not exceeding 150mm girth	m	0.10	1.50	0.30	0.27	2.07
150 to 300mm girth	m	0.20	3.00	0.40	0.51	3.91

skirtings and rails

over 300mm girth	m2	0.54	8.10	1.70	1.47	11.27
not exceeding 150mm girth	m	0.10	1.50	0.30	0.27	2.07
150 to 300mm girth	m	0.20	3.00	0.40	0.51	3.91

One coat acrylic primer,
one coloured oil-based undercoat
and one coloured eggshell on
on wood

general surfaces

over 300mm girth	m2	0.50	7.50	1.76	1.39	10.65
not exceeding 150mm girth	m	0.08	1.20	0.30	0.23	1.73

	Unit	Labour Hours	Mat'ls £	O & P £	Total £
150 to 300mm girth	m	0.18	2.70	0.62	0.50	3.82
isolated areas not						
exceeding 0.5m2	nr	0.25	3.75	0.88	0.69	5.32
windows, screens glazed						
doors and the like						
panes area not exceeding						
0.2m2	m2	1.01	15.15	1.12	2.44	18.71
panes area 0.1 to 0.5m2	m2	0.86	12.90	1.00	2.09	15.99
panes area 0.5 to 1m2	m2	0.71	10.65	0.86	1.73	13.24
panes area exceeding 1m2	m2	0.62	9.30	0.74	1.51	11.55
frames and linings						
over 300mm girth	m2	0.54	8.10	1.76	1.48	11.34
not exceeding 150mm						
girth	m	0.10	1.50	0.30	0.27	2.07
150 to 300mm girth	m	0.20	3.00	0.62	0.54	4.16
skirtings and rails						
over 300mm girth	m2	0.54	8.10	1.76	1.48	11.34
not exceeding 150mm						
girth	m	0.10	1.50	0.30	0.27	2.07
150 to 300mm girth	m	0.20	3.00	0.62	0.54	4.16

One coat zinc chromate
primer, one white oil-based
undercoat and one white
oil-based finishing coat on
metal

	Unit	Labour Hours	Mat'ls £	O & P £	Total £
general surfaces						
over 300mm girth	m2	0.50	7.50	1.60	1.37	10.47
not exceeding 150mm						
girth	m	0.08	1.20	0.26	0.22	1.68

	Unit	Labour Hours	£	Mat'ls £	O & P £	Total £
One coat zinc chromate primer, one white oil-based undercoat and one white oil-based finishing coat (cont'd)						
150 to 300mm girth	m	0.18	2.70	0.54	0.49	3.73
isolated areas not						
exceeding 0.5m2	nr	0.25	3.75	0.80	0.68	5.23
windows, screens glazed doors and the like						
panes area not exceeding						
0.2m2	m2	1.01	15.15	1.00	2.42	18.57
panes area 0.1 to 0.5m2	m2	0.86	12.90	0.86	2.06	15.82
panes area 0.5 to 1m2	m2	0.71	10.65	0.74	1.71	13.10
panes area exceeding 1m2	m2	0.62	9.30	0.60	1.49	11.39
structural metalwork, general surfaces						
over 300mm girth	m2	0.78	11.70	1.60	2.00	15.30
not exceeding 150mm girth	m	0.15	2.25	0.26	0.38	2.89
150 to 300mm girth	m	0.28	4.20	0.54	0.71	5.45
isolated areas not						
exceeding 0.5m2	nr	0.39	5.85	0.80	1.00	7.65
structural metalwork, roof truss members						
over 300mm girth	m2	1.20	18.00	1.60	2.94	22.54
not exceeding 150mm girth	m	0.24	3.60	0.26	0.58	4.44
150 to 300mm girth	m	0.18	2.70	0.54	0.49	3.73
isolated areas not						
exceeding 0.5m2	nr	0.60	9.00	0.80	1.47	11.27

	Unit	Labour	Hours £	Mat'ls £	O & P £	Total £
radiators, panel type						
over 300mm girth	m2	0.50	7.50	1.28	1.32	10.10
not exceeding 150mm girth	m	0.08	1.20	0.26	0.22	1.68
150 to 300mm girth	m	0.18	2.70	0.56	0.49	3.75
isolated areas not exceeding 0.5m2	nr	0.25	3.75	0.64	0.66	5.05
radiators, column type						
over 300mm girth	m2	0.76	11.40	1.28	1.90	14.58
not exceeding 150mm girth	m	0.15	2.25	0.26	0.38	2.89
150 to 300mm girth	m	0.29	4.35	0.56	0.74	5.65
isolated areas not exceeding 0.5m2	nr	0.38	5.70	0.64	0.95	7.29

One coat zinc chromate
primer, one coloured oil-based
undercoat and one coloured
oil-based finishing coat on
metal

	Unit	Labour	Hours £	Mat'ls £	O & P £	Total £
general surfaces						
over 300mm girth	m2	0.50	7.50	2.08	1.44	11.02
not exceeding 150mm girth	m	0.08	1.20	0.28	0.22	1.70
150 to 300mm girth	m	0.18	2.70	0.58	0.49	3.77
isolated areas not exceeding 0.5m2	nr	0.25	3.75	0.84	0.69	5.28
windows, screens glazed doors and the like						
panes area not exceeding 0.2m2	m2	1.01	15.15	1.02	2.43	18.60

	Unit	Labour Hours	£ Labour	Mat'ls £	O & P £	Total £
One coat zinc chromate primer, one coloured oil-based undercoat and one coloured oil-based finishing coat (cont'd)						
panes area 0.1 to 0.5m2	m2	0.86	12.90	0.90	2.07	15.87
panes area 0.5 to 1m2	m2	0.71	10.65	0.78	1.71	13.14
panes area exceeding 1m2	m2	0.62	9.30	0.58	1.48	11.36
structural metalwork, general surfaces						
over 300mm girth	m2	0.78	11.70	1.76	2.02	15.48
not exceeding 150mm girth	m	0.15	2.25	0.28	0.38	2.91
150 to 300mm girth	m	0.28	4.20	0.60	0.72	5.52
isolated areas not exceeding 0.5m2	nr	0.39	5.85	0.84	1.00	7.69
structural metalwork, roof truss members						
over 300mm girth	m2	1.20	18.00	1.72	2.96	22.68
not exceeding 150mm girth	m	0.24	3.60	0.28	0.58	4.46
150 to 300mm girth	m	0.40	6.00	0.60	0.99	7.59
isolated areas not exceeding 0.5m2	nr	0.60	9.00	0.84	1.48	11.32
radiators, panel type						
over 300mm girth	m2	0.50	7.50	1.72	1.38	10.60
not exceeding 150mm girth	m	0.08	1.20	0.28	0.22	1.70
150 to 300mm girth	m	0.18	2.70	0.60	0.50	3.80
isolated areas not exceeding 0.5m2	nr	0.25	3.75	0.84	0.69	5.28

	Unit	Labour Hours £	Mat'ls £	O & P £	Total £
radiators, column type					
over 300mm girth	m2	0.76	11.40	1.72	1.97	15.09
not exceeding 150mm					
girth	m	0.15	2.25	0.28	0.38	2.91
150 to 300mm girth	m	0.29	4.35	0.56	0.74	5.65
isolated areas not					
exceeding 0.5m2	nr	0.38	5.70	0.66	0.95	7.31

One coat red oxide primer,
one white oil-based
undercoat and one white
oil-based finishing coat on
metal

	Unit	Labour Hours £	Mat'ls £	O & P £	Total £
general surfaces					
over 300mm girth	m2	0.50	7.50	1.54	1.36	10.40
not exceeding 150mm					
girth	m	0.08	1.20	0.25	0.22	1.67
150 to 300mm girth	m	0.18	2.70	0.53	0.48	3.71
isolated areas not					
exceeding 0.5m2	nr	0.25	3.75	0.76	0.68	5.19
windows, screens glazed					
doors and the like					
panes area not exceeding					
0.2m2	m2	1.01	15.15	0.98	2.42	18.55
panes area 0.1 to 0.5m2	m2	0.86	12.90	0.84	2.06	15.80
panes area 0.5 to 1m2	m2	0.71	10.65	0.72	1.71	13.08
panes area exceeding 1m2	m2	0.62	9.30	0.58	1.48	11.36
structural metalwork, general					
surfaces					
over 300mm girth	m2	0.78	11.70	1.54	1.99	15.23

	Unit	Labour Hours £	Mat'ls £	O & P £	Total £

One coat red oxide primer, one white oil-based undercoat and one white oil-based finishing coat (cont'd)

	Unit	Labour Hours	Mat'ls	O & P	Total
not exceeding 150mm girth	m	0.15	2.25	0.23	0.37	2.85
150 to 300mm girth	m	0.28	4.20	0.53	0.71	5.44
isolated areas not exceeding 0.5m2	nr	0.39	5.85	0.76	0.99	7.60
structural metalwork, roof truss members						
over 300mm girth	m2	1.20	18.00	1.54	2.93	22.47
not exceeding 150mm girth	m	0.24	3.60	0.25	0.58	4.43
150 to 300mm girth	m	0.40	6.00	0.53	0.98	7.51
isolated areas not exceeding 0.5m2	nr	0.60	9.00	0.76	1.46	11.22
radiators, panel type						
over 300mm girth	m2	0.50	7.50	1.54	1.36	10.40
not exceeding 150mm girth	m	0.08	1.20	0.25	0.22	1.67
150 to 300mm girth	m	0.18	2.70	0.53	0.48	3.71
isolated areas not exceeding 0.5m2	nr	0.25	3.75	0.76	0.68	5.19
radiators, column type						
over 300mm girth	m2	0.76	11.40	1.54	1.94	14.88
not exceeding 150mm girth	m	0.20	3.00	0.25	0.49	3.74
150 to 300mm girth	m	0.24	3.60	0.53	0.62	4.75

	Unit	Labour	Hours £	Mat'ls £	O & P £	Total £
isolated areas not exceeding 0.5m2	nr	0.38	5.70	0.76	0.97	7.43

Two coats standard colour rubber paint on surfaces over 300mm girth

	Unit	Labour	Hours £	Mat'ls £	O & P £	Total £
brickwork walls	m2	0.40	6.00	5.70	1.76	13.46
brickwork walls in staircase areas	m2	0.44	6.60	5.70	1.85	14.15
blockwork walls	m2	0.44	6.60	6.24	1.93	14.77
blockwork walls in staircase areas	m2	0.48	7.20	6.24	2.02	15.46
concrete walls	m2	0.36	5.40	5.28	1.60	12.28
concrete walls in staircase areas	m2	0.40	6.00	5.28	1.69	12.97
concrete ceilings	m2	0.44	6.60	5.28	1.78	13.66
concrete ceilings in staircase areas	m2	0.48	7.20	5.28	1.87	14.35
plastered walls	m2	0.36	5.40	5.28	1.60	12.28
plastered walls in staircase areas	m2	0.40	6.00	5.28	1.69	12.97
plastered ceilings	m2	0.44	6.60	5.28	1.78	13.66
plastered ceilings in staircase areas	m2	0.48	7.20	5.28	1.87	14.35

Two coats rich colour rubber paint on surfaces over 300mm girth

	Unit	Labour	Hours £	Mat'ls £	O & P £	Total £
brickwork walls	m2	0.40	6.00	6.44	1.87	14.31
brickwork walls in staircase areas	m2	0.44	6.60	6.44	1.96	15.00
blockwork walls	m2	0.44	6.60	6.44	1.96	15.00

	Unit	Labour	Hours £	Mat'ls £	O & P £	Total £

**Two coats rich colour
rubber paint (cont'd)**

	Unit	Labour	Hours	Mat'ls	O & P	Total
blockwork walls in staircase						
areas	m2	0.48	7.20	6.44	2.05	15.69
concrete walls	m2	0.36	5.40	6.44	1.78	13.62
concrete walls in staircase						
areas	m2	0.40	6.00	6.44	1.87	14.31
concrete ceilings	m2	0.44	6.60	6.44	1.96	15.00
concrete ceilings in						
staircase areas	m2	0.48	7.20	6.44	2.05	15.69
plastered walls	m2	0.36	5.40	6.44	1.78	13.62
plastered walls in staircase						
areas	m2	0.40	6.00	6.44	1.87	14.31
plastered ceilings	m2	0.44	6.60	6.44	1.96	15.00
plastered ceilings in staircase						
areas	m2	0.48	7.20	6.44	2.05	15.69

**One coat Artex sealer/under-
coat and one Artex finishing
coat on surfaces over 300mm
girth**

	Unit	Labour	Hours	Mat'ls	O & P	Total
brickwork walls	m2	0.40	6.00	1.42	1.11	8.53
brickwork walls in staircase						
areas	m2	0.44	6.60	1.42	1.20	9.22
blockwork walls	m2	0.44	6.60	1.42	1.20	9.22
blockwork walls in staircase						
areas	m2	0.48	7.20	1.42	1.29	9.91
concrete walls	m2	0.36	5.40	1.42	1.02	7.84
concrete walls in staircase						
areas	m2	0.40	6.00	1.42	1.11	8.53
concrete ceilings	m2	0.44	6.60	1.42	1.20	9.22
concrete ceilings in						
staircase areas	m2	0.48	7.20	1.42	1.29	9.91

	Unit	Labour	Hours £	Mat'ls £	O & P £	Total £
plastered walls	m2	0.36	5.40	1.42	1.02	7.84
plastered walls in staircase areas	m2	0.40	6.00	1.42	1.11	8.53
plastered ceilings	m2	0.44	6.60	1.42	1.20	9.22
plastered ceilings in staircase areas	m2	0.48	7.20	1.42	1.29	9.91

Two coats clear polyurethane varnish on surfaces over 300mm girth

	Unit	Labour	Hours £	Mat'ls £	O & P £	Total £
general surfaces						
over 300mm girth	m2	0.32	4.80	3.52	1.25	9.57
not exceeding 150mm girth	m	0.05	0.75	0.58	0.20	1.53
150 to 300mm girth	m	0.10	1.50	1.16	0.40	3.06
isolated areas not exceeding 0.5m2	nr	0.16	2.40	1.76	0.62	4.78
windows, screens glazed doors and the like						
panes area not exceeding 0.2m2	m2	0.66	9.90	3.24	1.97	15.11
panes area 0.1 to 0.5m2	m2	0.56	8.40	3.16	1.73	13.29
panes area 0.5 to 1m2	m2	0.46	6.90	3.06	1.49	11.45
panes area exceeding 1m2	m2	0.40	6.00	3.00	1.35	10.35
frames and linings						
over 300mm girth	m2	0.32	4.80	3.52	1.25	9.57
not exceeding 150mm girth	m	0.05	0.75	0.60	0.20	1.55
150 to 300mm girth	m	0.10	1.50	1.16	0.40	3.06

	Unit	Labour Hours £	Mat'ls £	O & P £	Total £
Two coats clear polyurethane varnish (cont'd)					
skirtings and rails					
over 300mm girth	m2	0.32	4.80	3.52	1.25	9.57
not exceeding 150mm girth	m	0.05	0.75	0.60	0.20	1.55
150 to 300mm girth	m	0.10	1.50	1.16	0.40	3.06
Two coats coloured polyurethane varnish on surfaces over 300mm girth					
general surfaces					
over 300mm girth	m2	0.32	4.80	4.00	1.32	10.12
not exceeding 150mm girth	m	0.05	0.75	0.66	0.21	1.62
150 to 300mm girth	m	0.10	1.50	1.36	0.43	3.29
isolated areas not exceeding 0.5m2	nr	0.16	2.40	2.00	0.66	5.06
windows, screens glazed doors and the like					
panes area not exceeding 0.2m2	m2	0.66	9.90	3.54	2.02	15.46
panes area 0.1 to 0.5m2	m2	0.56	8.40	3.44	1.78	13.62
panes area 0.5 to 1m2	m2	0.46	6.90	3.66	1.58	12.14
panes area exceeding 1m2	m2	0.40	6.00	3.28	1.39	10.67
frames and linings					
over 300mm girth	m2	0.32	4.80	4.00	1.32	10.12
not exceeding 150mm girth	m	0.05	0.75	0.66	0.21	1.62
150 to 300mm girth	m	0.10	1.50	1.36	0.43	3.29

	Unit	Labour Hours	Mat'ls £	O & P £	Total £
skirtings and rails					
over 300mm girth	m2	0.32	4.80	4.00	1.32	10.12
not exceeding 150mm girth	m	0.05	0.75	0.66	0.21	1.62
150 to 300mm girth	m	0.10	1.50	1.36	0.43	3.29

EXTERNALLY

One coat wood primer, one white oil-based undercoat and one white oil-based finishing coat on wood

general surfaces					
over 300mm girth	m2	0.58	8.70	0.70	0.54	4.13
not exceeding 150mm girth	m	0.10	1.50	0.30	0.15	1.18
150 to 300mm girth	m	0.22	3.30	0.62	0.26	1.97
isolated areas not exceeding 0.5m2	nr	0.29	4.35	0.88	0.29	2.21

windows, screens glazed doors and the like

panes area not exceeding 0.2m2	m2	1.09	16.35	1.10	0.82	6.27
panes area 0.1 to 0.5m2	m2	0.94	14.10	0.96	0.71	5.47
panes area 0.5 to 1m2	m2	0.79	11.85	0.84	0.61	4.68
panes area exceeding 1m2	m2	0.70	10.50	0.72	0.55	4.19

frames and linings					
over 300mm girth	m2	0.60	9.00	1.76	0.69	5.32
not exceeding 150mm girth	m	0.11	1.65	0.30	0.15	1.18
150 to 300mm girth	m	0.23	3.45	0.62	0.26	1.97

	Unit	Labour Hours	Mat'ls £	O & P £	Total £

One coat wood primer, one white oil-based undercoat and one white oil-based finishing coat (cont'd)

railings, fences and gates
plain open type

over 300mm girth	m2	0.60	9.00	1.68	0.69	5.32
not exceeding 150mm girth	m	0.11	1.65	0.28	0.15	1.18
150 to 300mm girth	m	0.23	3.45	0.62	0.26	1.97

railings, fences and gates
ornamental type

over 300mm girth	m2	0.64	9.60	1.68	0.62	4.73
not exceeding 150mm girth	m	0.12	1.80	0.28	0.14	1.04
150 to 300mm girth	m	0.24	3.60	0.62	0.24	1.82

One coat wood primer, one coloured oil-based undercoat and one coloured oil-based finishing coat on wood

general surfaces

over 300mm girth	m2	0.58	8.70	1.92	1.59	12.21
not exceeding 150mm girth	m	0.10	1.50	0.32	0.27	2.09
150 to 300mm girth	m	0.22	3.30	0.66	0.59	4.55
isolated areas not exceeding 0.5m2	nr	0.29	4.35	0.94	0.79	6.08

	Unit	Labour Hours	Mat'ls £	O & P £	Total £
windows, screens glazed doors and the like					
panes area not exceeding 0.2m2	m2	1.09	16.35	1.14	2.62	20.11
panes area 0.1 to 0.5m2	m2	0.94	14.10	1.00	2.27	17.37
panes area 0.5 to 1m2	m2	0.79	11.85	0.88	1.91	14.64
panes area exceeding 1m2	m2	0.70	10.50	0.74	1.69	12.93
frames and linings					
over 300mm girth	m2	0.60	9.00	1.82	1.62	12.44
not exceeding 150mm girth	m	0.11	1.65	0.24	0.28	2.17
150 to 300mm girth	m	0.23	3.45	0.68	0.62	4.75
rails					
over 300mm girth	m2	0.60	9.00	1.82	1.62	12.44
not exceeding 150mm girth	m	0.11	1.65	0.24	0.28	2.17
150 to 300mm girth	m	0.23	3.45	0.68	0.62	4.75
One coat aluminium primer, one white oil-based undercoat and one white oil-based finishing coat on wood					
general surfaces					
over 300mm girth	m2	0.58	8.70	2.40	1.67	12.77
not exceeding 150mm girth	m	0.10	1.50	0.40	0.29	2.19
150 to 300mm girth	m	0.22	3.30	0.82	0.62	4.74
isolated areas not exceeding 0.5m2	nr	0.29	4.35	1.22	0.84	6.41

	Unit	Labour	Hours £	Mat'ls £	O & P £	Total £

**One coat aluminium primer,
one white oil-based undercoat
and one white oil-based
finishing coat (cont'd)**

windows, screens glazed
doors and the like

	Unit	Labour	Hours £	Mat'ls £	O & P £	Total £
panes area not exceeding 0.2m2	m2	1.05	15.75	1.40	2.57	19.72
panes area 0.1 to 0.5m2	m2	0.90	13.50	1.28	2.22	17.00
panes area 0.5 to 1m2	m2	0.75	11.25	1.02	1.84	14.11
panes area exceeding 1m2	m2	0.66	9.90	0.60	1.58	12.08
frames and linings						
over 300mm girth	m2	0.60	9.00	2.40	1.71	13.11
not exceeding 150mm girth	m	0.11	1.65	0.40	0.31	2.36
150 to 300mm girth	m	0.23	3.45	0.82	0.64	4.91
rails						
over 300mm girth	m2	0.80	12.00	2.40	2.16	16.56
not exceeding 150mm girth	m	0.12	1.80	0.40	0.33	2.53
150 to 300mm girth	m	0.40	6.00	0.82	1.02	7.84

**One coat aluminium primer,
one coloured oil-based undercoat
and one coloured oil-based
finishing coat on wood**

general surfaces

	Unit	Labour	Hours £	Mat'ls £	O & P £	Total £
over 300mm girth	m2	0.58	8.70	2.44	1.67	12.81
not exceeding 150mm girth	m	0.10	1.50	0.42	0.29	2.21

	Unit	Labour	Hours £	Mat'ls £	O & P £	Total £
150 to 300mm girth isolated areas not	m	0.22	3.30	0.84	0.62	4.76
exceeding 0.5m2	nr	0.29	4.35	1.24	0.84	6.43
windows, screens glazed doors and the like						
panes area not exceeding 0.2m2	m2	1.09	16.35	1.44	2.67	20.46
panes area 0.1 to 0.5m2	m2	0.94	14.10	1.30	2.31	17.71
panes area 0.5 to 1m2	m2	0.79	11.85	1.16	1.95	14.96
panes area exceeding 1m2	m2	0.70	10.50	1.04	1.73	13.27
frames and linings						
over 300mm girth	m2	0.60	9.00	2.44	1.72	13.16
not exceeding 150mm girth	m	0.11	1.65	0.42	0.31	2.38
150 to 300mm girth	m	0.23	3.45	0.78	0.63	4.86
rails						
over 300mm girth	m2	0.60	9.00	2.44	1.72	13.16
not exceeding 150mm girth	m	0.11	1.65	0.42	0.31	2.38
150 to 300mm girth	m	0.23	3.45	0.78	0.63	4.86
railings, fences and gates plain open type						
over 300mm girth	m2	0.60	9.00	1.88	0.71	5.41
not exceeding 150mm girth	m	0.11	1.65	0.40	0.17	1.31
150 to 300mm girth	m	0.23	3.45	1.00	0.28	2.15
railings, fences and gates ornamental type						
over 300mm girth	m2	0.64	9.60	2.30	0.71	5.41
not exceeding 150mm girth	m	0.12	1.80	0.40	0.15	1.16

	Unit	Labour	Hours £	Mat'ls £	O & P £	Total £
One coat acrylic primer, one white oil-based undercoat and one white oil-based finishing coat on wood						
general surfaces						
over 300mm girth	m2	0.58	8.70	1.76	1.57	12.03
not exceeding 150mm girth	m	0.10	1.50	0.40	0.29	2.19
150 to 300mm girth	m	0.22	3.30	0.66	0.59	4.55
isolated areas not exceeding 0.5m2	nr	0.29	4.35	0.88	0.78	6.01
windows, screens glazed doors and the like						
panes area not exceeding 0.2m2	m2	1.35	20.25	1.14	3.21	24.60
panes area 0.1 to 0.5m2	m2	0.90	13.50	0.96	2.17	16.63
panes area 0.5 to 1m2	m2	0.75	11.25	0.84	1.81	13.90
panes area exceeding 1m2	m2	0.66	9.90	0.72	1.59	12.21
frames and linings						
over 300mm girth	m2	0.60	9.00	1.76	1.61	12.37
not exceeding 150mm girth	m	0.11	1.65	0.40	0.31	2.36
150 to 300mm girth	m	0.23	3.45	0.60	0.61	4.66
rails						
over 300mm girth	m2	0.60	9.00	1.76	1.61	12.37
not exceeding 150mm girth	m	0.11	1.65	0.40	0.31	2.36
150 to 300mm girth	m	0.23	3.45	0.60	0.61	4.66

	Unit	Labour Hours	£	Mat'ls £	O & P £	Total £

One coat acrylic primer,
one coloured oil-based undercoat
and one coloured oil-based
finishing coat on wood

	Unit	Labour	Hours £	Mat'ls £	O & P £	Total £
general surfaces						
over 300mm girth	m2	0.58	8.70	1.80	1.58	12.08
not exceeding 150mm girth	m	0.10	1.50	0.30	0.27	2.07
150 to 300mm girth	m	0.22	3.30	0.66	0.59	4.55
isolated areas not exceeding 0.5m2	nr	0.29	4.35	0.90	0.79	6.04
windows, screens glazed doors and the like						
panes area not exceeding 0.2m2	m2	1.09	16.35	1.12	2.62	20.09
panes area 0.1 to 0.5m2	m2	0.94	14.10	1.00	2.27	17.37
panes area 0.5 to 1m2	m2	0.79	11.85	0.86	1.91	14.62
panes area exceeding 1m2	m2	0.70	10.50	0.74	1.69	12.93
frames and linings						
over 300mm girth	m2	0.60	9.00	1.80	1.62	12.42
not exceeding 150mm girth	m	0.11	1.65	0.30	0.29	2.24
150 to 300mm girth	m	0.23	3.45	0.66	0.62	4.73
rails						
over 300mm girth	m2	0.60	9.00	1.80	1.62	12.42
not exceeding 150mm girth	m	0.11	1.65	0.30	0.29	2.24
150 to 300mm girth	m	0.23	3.45	0.66	0.62	4.73

	Unit	Labour Hours	Mat'ls £	O & P £	Total £

One coat zinc chromate
primer, one white oil-based
undercoat and one white
oil-based finishing coat on metal

	Unit	Labour Hours	Mat'ls £	O & P £	Total £
general surfaces					
over 300mm girth	m2	0.58	8.70	1.60	1.55	11.85
not exceeding 150mm girth	m	0.10	1.50	0.26	0.26	2.02
150 to 300mm girth	m	0.22	3.30	0.54	0.58	4.42
isolated areas not exceeding 0.5m2	nr	0.29	4.35	0.80	0.77	5.92
windows, screens glazed doors and the like					
panes area not exceeding 0.2m2	m2	1.09	16.35	1.00	2.60	19.95
panes area 0.1 to 0.5m2	m2	0.94	14.10	0.86	2.24	17.20
panes area 0.5 to 1m2	m2	0.79	11.85	0.74	1.89	14.48
panes area exceeding 1m2	m2	0.70	10.50	0.60	1.67	12.77
structural metalwork, general surfaces					
over 300mm girth	m2	0.60	9.00	1.60	1.59	12.19
not exceeding 150mm girth	m	0.11	1.65	0.26	0.29	2.20
150 to 300mm girth	m	0.23	3.45	0.54	0.60	4.59
isolated areas not exceeding 0.5m2	nr	0.34	5.10	0.80	0.89	6.79
eaves gutters					
over 300mm girth	m2	0.70	10.50	1.28	1.77	13.55
not exceeding 300mm girth	m	0.25	3.75	0.44	0.63	4.82
isolated areas not exceeding 0.5m2	nr	0.39	5.85	0.74	0.99	7.58

	Unit	Labour	Hours £	Mat'ls £	O & P £	Total £
services, pipes, conduits ducting and the like						
over 300mm girth	m2	0.66	9.90	1.28	1.68	12.86
not exceeding 300mm girth	m	0.22	3.30	0.44	0.56	4.30
isolated areas not exceeding 0.5m2	nr	0.33	4.95	0.74	0.85	6.54

One coat zinc chromate primer, one coloured oil-based undercoat and one coloured oil-based finishing coat on metal

	Unit	Labour	Hours £	Mat'ls £	O & P £	Total £
general surfaces						
over 300mm girth	m2	0.58	8.70	2.08	1.62	12.40
not exceeding 150mm girth	m	0.10	1.50	0.28	0.27	2.05
150 to 300mm girth	m	0.22	3.30	0.58	0.58	4.46
isolated areas not exceeding 0.5m2	nr	0.29	4.35	0.84	0.78	5.97
windows, screens glazed doors and the like						
panes area not exceeding 0.2m2	m2	1.13	16.95	1.02	2.70	20.67
panes area 0.1 to 0.5m2	m2	0.98	14.70	0.90	2.34	17.94
panes area 0.5 to 1m2	m2	0.83	12.45	0.78	1.98	15.21
panes area exceeding 1m2	m2	0.74	11.10	0.58	1.75	13.43
structural metalwork, general surfaces						
over 300mm girth	m2	0.60	9.00	2.08	1.66	12.74
not exceeding 150mm girth	m	0.11	1.65	0.28	0.29	2.22
150 to 300mm girth	m	0.23	3.45	0.58	0.60	4.63

	Unit	Labour	Hours £	Mat'ls £	O & P £	Total £

One coat zinc chromate primer, one coloured oil-based undercoat and one coloued oil-based finishing coat (cont'd)

	Unit	Labour	Hours £	Mat'ls £	O & P £	Total £
isolated areas not exceeding 0.5m2	nr	0.34	5.10	0.84	0.89	6.83
eaves gutters						
over 300mm girth	m2	0.60	9.00	1.56	1.58	12.14
not exceeding 300mm girth	m	0.11	1.65	0.56	0.33	2.54
isolated areas not exceeding 0.5m2	nr	0.30	4.50	0.78	0.79	6.07
services, pipes, conduits ducting and the like						
over 300mm girth	m2	0.60	9.00	1.66	1.60	12.26
not exceeding 300mm girth	m	0.11	1.65	0.56	0.33	2.54
isolated areas not exceeding 0.5m2	nr	0.30	4.50	0.76	0.79	6.05

One coat metal red oxide primer, one white oil-based undercoat and one white oil-based finishing coat on metal

	Unit	Labour	Hours £	Mat'ls £	O & P £	Total £
general surfaces						
over 300mm girth	m2	0.58	8.70	1.54	1.54	11.78
not exceeding 150mm girth	m	0.10	1.50	0.25	0.26	2.01
150 to 300mm girth	m	0.22	3.30	0.53	0.57	4.40
isolated areas not exceeding 0.5m2	nr	0.29	4.35	0.76	0.77	5.88

	Unit	Labour Hours	Mat'ls £	O & P £	Total £
windows, screens glazed					
doors and the like					
panes area not exceeding					
0.2m2	m2	1.09	16.35	0.98	2.60	19.93
panes area 0.1 to 0.5m2	m2	0.94	14.10	0.84	2.24	17.18
panes area 0.5 to 1m2	m2	0.79	11.85	0.72	1.89	14.46
panes area exceeding 1m2	m2	0.70	10.50	0.58	1.66	12.74
structural metalwork, general					
surfaces					
over 300mm girth	m2	0.60	9.00	1.54	1.58	12.12
not exceeding 150mm					
girth	m	0.11	1.65	0.25	0.29	2.19
150 to 300mm girth	m	0.23	3.45	0.53	0.60	4.58
isolated areas not					
exceeding 0.5m2	nr	0.30	4.50	0.76	0.79	6.05
eaves gutters					
over 300mm girth	m2	0.60	9.00	1.44	1.57	12.01
not exceeding 300mm					
girth	m	0.11	1.65	0.40	0.31	2.36
isolated areas not					
exceeding 0.5m2	nr	0.30	4.50	0.72	0.78	6.00
services, pipes, conduits					
ducting and the like					
over 300mm girth	m2	0.60	9.00	1.44	1.57	12.01
not exceeding 300mm					
girth	m	0.11	1.65	0.40	0.31	2.36
isolated areas not					
exceeding 0.5m2	nr	0.30	4.50	0.72	0.78	6.00

	Unit	Labour Hours	Mat'ls £	O & P £	Total £

One coat red oxide primer,
one coloured oil-based
undercoat and one coloued
oil-based finishing coat on
metal

general surfaces					
over 300mm girth	m2	0.60	9.00	1.62	1.59	12.21
not exceeding 150mm girth	m	0.11	1.65	0.28	0.29	2.22
150 to 300mm girth	m	0.23	3.45	0.56	0.60	4.61
isolated areas not exceeding 0.5m2	nr	0.30	4.50	0.82	0.80	6.12
windows, screens glazed doors and the like					
panes area not exceeding 0.2m2	m2	1.16	17.40	1.00	2.76	21.16
panes area 0.1 to 0.5m2	m2	0.96	14.40	0.88	2.29	17.57
panes area 0.5 to 1m2	m2	0.83	12.45	0.76	1.98	15.19
panes area exceeding 1m2	m2	0.74	11.10	0.56	1.75	13.41
structural metalwork, general surfaces					
over 300mm girth	m2	0.60	9.00	1.62	1.59	12.21
not exceeding 150mm girth	m	0.11	1.65	0.28	0.29	2.22
150 to 300mm girth	m	0.23	3.45	0.56	0.60	4.61
isolated areas not exceeding 0.5m2	nr	0.30	4.50	0.82	0.80	6.12
eaves gutters					
over 300mm girth	m2	0.60	9.00	1.60	1.59	12.19
not exceeding 300mm girth	m	0.11	1.65	0.54	0.33	2.52

	Unit	Labour	Hours £	Mat'ls £	O & P £	Total £
isolated areas not exceeding 0.5m2	nr	0.30	4.50	0.76	0.79	6.05
services, pipes, conduits ducting and the like						
over 300mm girth	m2	0.60	9.00	1.60	1.59	12.19
not exceeding 300mm girth	m	0.11	1.65	0.54	0.33	2.52
isolated areas not exceeding 0.5m2	nr	0.30	4.50	0.76	0.79	6.05

One coat masonry sealer, two coats Snowcem on surfaces over 300mm girth

	Unit	Labour	Hours £	Mat'ls £	O & P £	Total £
brickwork walls	m2	0.36	5.40	2.72	1.22	9.34
blockwork walls	m2	0.42	6.30	2.72	1.35	10.37
concrete walls	m2	0.33	4.95	2.42	1.11	8.48
rendered walls	m2	0.36	5.40	2.42	1.17	8.99
roughcast walls	m2	0.42	6.30	3.00	1.40	10.70

One coat masonry sealer, two coats Sandtex Matt on surfaces over 300mm girth

	Unit	Labour	Hours £	Mat'ls £	O & P £	Total £
brickwork walls	m2	0.36	5.40	4.00	1.41	10.81
blockwork walls	m2	0.42	6.30	4.00	1.55	11.85
concrete walls	m2	0.33	4.95	3.60	1.28	9.83
rendered walls	m2	0.36	5.40	3.60	1.35	10.35
roughcast walls	m2	0.42	6.30	4.40	1.61	12.31

	Unit	Labour Hours	Mat'ls £	O & P £	Total £

One coat masonry sealer, two
coats Sandtex Textured on surfaces
over 300mm girth

	Unit	Labour Hours	Mat'ls £	O & P £	Total £
brickwork walls	m2	0.36	5.40	8.90	2.15	16.45
blockwork walls	m2	0.42	6.30	8.90	2.28	17.48
concrete walls	m2	0.33	4.95	8.30	1.99	15.24
rendered walls	m2	0.36	5.40	8.30	2.06	15.76
roughcast walls	m2	0.42	6.30	9.05	2.30	17.65

Two coats clear preserver
on sawn wood surfaces

over 300mm girth	m2	0.36	5.40	1.92	1.10	8.42
not exceeding 300mm girth	m	0.12	1.80	0.60	0.36	2.76
isolated areas not exceeding 0.5m2	nr	0.18	2.70	0.96	0.55	4.21

railings, fences and gates
plain open type

over 300mm girth	m2	0.40	6.00	1.92	1.19	9.11
not exceeding 150mm girth	m	0.07	1.05	0.60	0.25	1.90
150 to 300mm girth	m	0.14	2.10	0.96	0.46	3.52

railings, fences and gates
ornamental type

over 300mm girth	m2	0.42	6.30	1.92	1.23	9.45
not exceeding 150mm girth	m	0.08	1.20	0.60	0.27	2.07
150 to 300mm girth	m	0.15	2.25	0.96	0.48	3.69

	Unit	Labour Hours	Mat'ls	O & P	Total
		£	£	£	£
Two coats clear preserver on wrot wood surfaces					
over 300mm girth	m2	0.36	5.40	1.86	1.09	8.35
not exceeding 300mm girth	m	0.12	1.80	0.58	0.36	2.74
isolated areas not exceeding 0.5m2	nr	0.18	2.70	0.90	0.54	4.14
railings, fences and gates plain open type					
over 300mm girth	m2	0.40	6.00	1.86	1.18	9.04
not exceeding 150mm girth	m	0.07	1.05	0.58	0.24	1.87
150 to 300mm girth	m	0.14	2.10	0.90	0.45	3.45
railings, fences and gates ornamental type					
over 300mm girth	m2	0.42	6.30	1.86	1.22	9.38
not exceeding 150mm girth	m	0.08	1.20	0.58	0.27	2.05
150 to 300mm girth	m	0.15	2.25	0.90	0.47	3.62
Two coats light preserver on sawn wood surfaces					
over 300mm girth	m2	0.36	5.40	1.90	1.10	8.40
not exceeding 300mm girth	m	0.12	1.80	0.60	0.36	2.76
isolated areas not exceeding 0.5m2	nr	0.18	2.70	0.96	0.55	4.21

	Unit	Labour	Hours £	Mat'ls £	O & P £	Total £
Two coats light preserver (cont'd)						
railings, fences and gates plain open type						
over 300mm girth	m2	0.40	6.00	1.90	1.19	9.09
not exceeding 150mm girth	m	0.07	1.05	0.60	0.25	1.90
150 to 300mm girth	m	0.14	2.10	0.96	0.46	3.52
railings, fences and gates ornamental type						
over 300mm girth	m2	0.42	6.30	1.90	1.23	9.43
not exceeding 150mm girth	m	0.08	1.20	0.60	0.27	2.07
150 to 300mm girth	m	0.15	2.25	0.96	0.48	3.69
Two coats light preserver on wrot wood surfaces						
over 300mm girth	m2	0.36	5.40	1.76	1.07	8.23
not exceeding 300mm girth	m	0.12	1.80	0.48	0.34	2.62
isolated areas not exceeding 0.5m2	nr	0.18	2.70	0.88	0.54	4.12
railings, fences and gates plain open type						
over 300mm girth	m2	0.40	6.00	1.76	1.16	8.92
not exceeding 150mm girth	m	0.07	1.05	0.48	0.23	1.76
150 to 300mm girth	m	0.14	2.10	0.88	0.45	3.43

	Unit	Labour	Hours £	Mat'ls £	O & P £	Total £
railings, fences and gates ornamental type						
over 300mm girth	m2	0.42	6.30	1.76	1.21	9.27
not exceeding 150mm girth	m	0.08	1.20	0.48	0.25	1.93
150 to 300mm girth	m	0.15	2.25	0.88	0.47	3.60
Two coats dark preserver on sawn wood surfaces						
over 300mm girth	m2	0.36	5.40	1.90	1.10	8.40
not exceeding 300mm girth	m	0.12	1.80	0.60	0.36	2.76
isolated areas not exceeding 0.5m2	nr	0.18	2.70	0.96	0.55	4.21
railings, fences and gates plain open type						
over 300mm girth	m2	0.40	6.00	1.90	1.19	9.09
not exceeding 150mm girth	m	0.07	1.05	0.60	0.25	1.90
150 to 300mm girth	m	0.14	2.10	0.96	0.46	3.52
railings, fences and gates ornamental type						
over 300mm girth	m2	0.42	6.30	1.90	1.23	9.43
not exceeding 150mm girth	m	0.08	1.20	1.76	0.44	3.40
150 to 300mm girth	m	0.15	2.25	0.48	0.41	3.14
Two coats dark preserver on wrot wood surfaces						
over 300mm girth	m2	0.36	5.40	1.76	1.07	8.23
not exceeding 300mm girth	m	0.12	1.80	0.48	0.34	2.62

	Unit	Labour Hours	£	Mat'ls £	O & P £	Total £

Two coats dark preserver (cont'd)

	Unit	Labour Hours	Mat'ls £	O & P £	Total £
isolated areas not exceeding 0.5m2	nr	0.18	2.70	0.88	0.54	4.12
railings, fences and gates plain open type						
over 300mm girth	m2	0.40	6.00	1.76	1.16	8.92
not exceeding 150mm girth	m	0.07	1.05	0.48	0.23	1.76
150 to 300mm girth	m	0.14	2.10	0.88	0.45	3.43
railings, fences and gates ornamental type						
over 300mm girth	m2	0.42	6.30	1.76	1.21	9.27
not exceeding 150mm girth	m	0.08	1.20	0.48	0.25	1.93
150 to 300mm girth	m	0.15	2.25	0.88	0.47	3.60

One coat green preserver on sawn wood surfaces

	Unit	Labour Hours	Mat'ls £	O & P £	Total £
over 300mm girth	m2	0.36	5.40	1.90	1.10	8.40
not exceeding 300mm girth	m	0.12	1.80	0.60	0.36	2.76
isolated areas not exceeding 0.5m2	nr	0.18	2.70	0.96	0.55	4.21
railings, fences and gates plain open type						
over 300mm girth	m2	0.40	6.00	1.90	1.19	9.09
not exceeding 150mm girth	m	0.07	1.05	0.60	0.25	1.90
150 to 300mm girth	m	0.14	2.10	0.96	0.46	3.52

	Unit	Labour	Hours £	Mat'ls £	O & P £	Total £
railings, fences and gates ornamental type						
over 300mm girth	m2	0.42	6.30	1.90	1.23	9.43
not exceeding 150mm girth	m	0.08	1.20	0.60	0.27	2.07
150 to 300mm girth	m	0.15	2.25	0.96	0.48	3.69
One coat green preserver on wrot wood surfaces						
over 300mm girth	m2	0.36	5.40	1.76	1.07	8.23
not exceeding 300mm girth	m	0.12	1.80	0.48	0.34	2.62
isolated areas not exceeding 0.5m2	nr	0.18	2.70	0.88	0.54	4.12
railings, fences and gates plain open type						
over 300mm girth	m2	0.40	6.00	1.90	1.19	9.09
not exceeding 150mm girth	m	0.07	1.05	0.60	0.25	1.90
150 to 300mm girth	m	0.14	2.10	0.96	0.46	3.52
railings, fences and gates ornamental type						
over 300mm girth	m2	0.42	6.30	1.76	1.21	9.27
not exceeding 150mm girth	m	0.08	1.20	0.48	0.25	1.93
150 to 300mm girth	m	0.15	2.25	0.88	0.47	3.60

Part Five

PROJECT COSTS – PAINTING AND WALLPAPERING

Project costs

PROJECT COSTS

This section gives the approximate costs of carrying out painting and
wallpapering in a typical house. The following assumptions have
been made.

Living room:	1 door, 1 fireplace, 1 large window
Dining room:	2 doors, 1 large window
Bedroom	1 door, 1 average window
Kitchen	2 doors, fittings below dado level, 1 large window
WC	1 door, 1 small window
Bathroom	1 door, 1 average window, 1 cupboard

Ceiling heights: Ground floor 2.4 metres and first floor 2.2 metres.

The following are typical rounded-off quantities for painting ceilings
and walls.

	Ceilings	Walls
Living room		
3.6 x 3.0m	11m2	24m2
3.6 x 3.6m	13m2	26m2
4.2 x 3.0m	13m2	26m2
4.2 x 3.6m	15m2	29m2
4.2 x 4.2m	18m2	32m2
4.8 x 3.0m	15m2	29m2
4.8 x 3.6m	17m2	32m2
4.8 x 4.2m	20m2	34m2
4.8 x 4.8m	23m2	37m2

Dining room

3.0 x 3.0m	9m2	20m2
3.6 x 3.0m	11m2	24m2
3.6 x 3.6m	13m2	26m2
4.2 x 3.0m	13m2	26m2
4.2 x 3.6m	15m2	29m2
4.2 x 4.2m	18m2	32m2
4.8 x 3.0m	15m2	29m2
4.8 x 3.6m	17m2	32m2
4.8 x 4.2m	20m2	24m2

	Ceilings	Walls

Bedroom

	Ceilings	Walls
2.4 x 2.4m	6m2	17m2
2.4 x 3.0m	7m2	20m2
2.4 x 3.6m	9m2	22m2
3.0 x 3.0m	9m2	22m2
3.0 x 3.6m	11m2	25m2
3.6 x 3.6m	13m2	27m2
3.6 x 4.2m	15m2	30m2
3.6 x 4.8m	17m2	32m2
4.2 x 4.2m	18m2	32m2
4.2 x 4.8m	20m2	35m2

Kitchen

	Ceilings	Walls
1.8 x 3.0m	6m2	7m2
1.8 x 3.6m	7m2	10m2
2.4 x 3.0m	7m2	10m2
2.4 x 3.6m	9m2	11m2
3.0 x 3.0m	9m2	11m2
3.0 x 3.6m	11m2	11m2
3.0 x 4.2m	13m2	11m2
3.6 x 3.6m	13m2	13m2
3.6 x 4.2m	15m2	15m2

WC

1.0 x 1.5m	2m2	8m2
1.2 x 1.5m	2m2	9m2
1.2 x 1.8m	2m2	10m2
1.3 x 1.5m	2m2	9m2
1.3 x 1.8m	2m2	10m2

Bathroom

1.8 x 2.4m	4m2	23m2
1.8 x 3.0m	5m2	25m2
2.1 x 3.4m	5m2	23m2
2.4 x 2.4m	6m2	25m2
2.4 x 3.0m	7m2	28m2
2.4 x 3.6m	9m2	30m2

The specification for the work carried out is:

Ceilings: Two coats matt white emulsion paint

Walls: Vinyl wallpaper £6.00 per roll

Woodwork: One coat wood primer, one coat white oil-based
undercoat and one coat oil-based finishing coat
on wood surfaces.

The cost of the work in each room is assessed by multiplying the cost
data in Unit Rates by the quantities listed above. Note that all the
figures are rounded off to the nearest pound.

	£
Living room	
3.6 x 3.0m	420
3.6 x 3.6m	460
4.2 x 3.0m	460
4.2 x 3.6m	535
4.2 x 4.2m	615
4.8 x 3.0m	535
4.8 x 3.6m	600
4.8 x 4.2m	665
4.8 x 4.8m	745

Dining room

3.0 x 3.0m	380
3.6 x 3.0m	420
3.6 x 3.6m	460
4.2 x 3.0m	460
4.2 x 3.6m	535
4.2 x 4.2m	615
4.8 x 3.0m	535
4.8 x 3.6m	600
4.8 x 4.2m	665

Bedroom

2.4 x 2.4m	270
2.4 x 3.0m	320
2.4 x 3.6m	370
3.0 x 3.0m	370
3.0 x 3.6m	435
3.6 x 3.6m	470
3.6 x 4.2m	550
3.6 x 4.8m	600
4.2 x 4.2m	615
4.2 x 4.8m	680

Kitchen

1.8 x 3.0m	170
1.8 x 3.6m	215
2.4 x 3.0m	215
2.4 x 3.6m	255
3.0 x 3.0m	255
3.0 x 3.6m	285
3.0 x 4.2m	305
3.6 x 3.6m	325
3.6 x 4.2m	390

WC

1.0 x 1.5m	115
1.2 x 1.5m	125
1.2 x 1.8m	135
1.3 x 1.5m	125
1.3 x 1.8m	135

Bathroom

1.8 x 2.4m	305
1.8 x 3.0m	340
2.1 x 2.4m	320
2.4 x 2.4m	355
2.4 x 3.0m	400
2.4 x 3.6m	455

Part Six

BUSINESS MATTERS

Starting a business

Running a business

Taxation

STARTING A BUSINESS

Most small businesses come into being for one of two reasons – ambition or desperation! A person with genuine ambition for commercial success will never be completely satisfied until he has become self-employed and started his own business. But many successful businesses have been started because the proprietor was forced into this course of action because of redundancy.

Before giving up his job, the would-be businessman should consider carefully whether he has the required skills and the temperament to survive in the highly competitive self-employed market. Before commencing in business it is essential to assess the commercial viability of the intended business because it is pointless to finance a business that is not going to be commercially viable.

In the early stages it is important to make decisions such as: What exactly is the product being sold? What is the market view of that product? What steps are required before the developed product is first sold and where are those sales coming from?

As much information as possible should be obtained on how to run a business before taking the plunge. Sales targets should be set and it should be clearly established how those important first sales are obtained. Above all, do not underestimate the amount of time required to establish and finance a new business venture.

Whatever the size of the business it is important that you put in writing exactly what you are trying to do. This means preparing a business plan that will not only assist in establishing your business aims but is essential if you need to raise finance. The contents of a typical business plan are set out later. It is important to realise that you are not on your own and there are many contacts and advising agencies that can be of assistance.

Potential customers and trade contacts

Many persons intending to start a business in the construction industry will have already had experience as employees. Use all contacts to check the market, establish the sort of work that is available and the current charge out rates.

In the domestic market, check on the competition for prices and services provided. Study advertisements for your kind of work and try to get firm promises of work before the start-up date.

Testing the market

Talk to as many traders as possible operating in the same field. Identify if the market is in the industrial, commercial, local government or in the domestic field. Talk to prospective customers and clients and consider how you can improve on what is being offered in terms of price, quality, speed, convenience, reliability and back-up service.

Business links

There is no shortage of information about the many aspects of starting and running your own business. Finance, marketing, legal requirements, developing your business idea and taxation matters are all the subject of a mountain of books, pamphlets, guides and courses so it should not be necessary to pay out a lot of money for this information. Indeed, the likelihood is that the aspiring businessman will be overwhelmed with information and will need professional guidance to reduce the risk of wasting time on studying unnecessary subjects.

Business Links are now well established and provide a good place to start for both information and advice. These organisations provide a 'one-stop-shop' for advice and assistance to owner-managed businesses. They will often replace the need to contact Training and Enterprise Councils (TECs) and many of the other official organisations listed below. Point of contact: telephone directory for address.

Training and Enterprise Councils (TECs)

TECs are comprised of a board of directors drawn from the top men in local industry, commerce, education, trade unions etc., who, together with their staff and experienced business counsellors, assist both new and established concerns in all aspects of running a business. This takes the form of across-the-table advice and also hands-on assistance in management, marketing and finance if required. There are also training courses and seminars available in most areas together with the possibility of grants in some areas.
Point of contact: local Jobcentre or Citizens' Advice Bureau.

Banks

Approach banks for information about the business accounts and financial services that are available. Your local Business Link can advise on how best to find a suitable bank manager and inform you as to what the bank will require.

Shop around several banks and branches if you are not satisfied at first because managers vary widely in their views on what is a viable business proposition. Remember, most banks have useful free information packs to help business start-up.

Point of contact: local bank manager.

HM Inspector of Taxes

Make a preliminary visit to the local tax office enquiry counter for their publications on income tax and national insurance contributions.

SA/Bk 3	Self assessment. A guide to keeping records for the self employed
IR 15(CIS)	Construction Industry Tax Deduction Scheme
CWL	Starting your own business
IR 40(CIS)	Conditions for Getting a Sub-Contractor's Tax Certificate
NE1	PAYE for Employers (if you employ someone)
NE3	PAYE for new and small Employers
IR 56/N139	Employed or Self-Employed. A guide for tax and National Insurance
CA02	National Insurance contributions for self employed people with small earnings.

Remember, the onus is on the taxpayer, within three months, to notify the Inland Revenue that he is in business and failure to do so may result in the imposition of £100 penalty. Either send a letter or use the form provided at the back of the *'Starting your own business booklet'* to the Inland Revenue National Insurance Contributions Office and they will inform your local tax office of the change in your employment status.

Point of contact: telephone directory for address.

Inland Revenue National Insurance Contributions Office

Self Employment Services
Customer Accounts Section
Longbenton
Newcastle NE 98 1ZZ

Telephone the Call Centre on 0845 9154655 and ask for the following publications:

CWL2	Class 2 and Class 4 Contributions for the Self Employed
CA02	People with Small Earnings from Self-Employment
CA04	Direct Debit – The Easy Way to Pay. Class 2 and Class 3
CA07	Unpaid and Late Paid Contributions and for Employers
CWG1	Employer's Quick Guide to PAYE and NIC Contributions
CA30	Employer's Manual to Statutory Sick Pay

VAT

The VAT office also offer a number of useful publications, including;

700	The VAT Guide
700/1	Should I be Registered for VAT?
731	Cash Accounting
732	Annual Accounting
742	Land and Property

Information about the Cash Accounting Scheme and the introduction of annual VAT returns are dealt with later.
Point of contact: telephone directory for address.

Local authorities

Authorities vary in provisions made for small businesses but all have been asked to simplify and cut delays in planning applications. In Assisted Areas, rent-free periods and reductions in rates may be available on certain industrial and commercial properties. As a preliminary to either purchasing or renting business premises, the following booklets will be helpful:

Step by Step Guide to Planning Permission for Small Businesses, and *Business Leases and Security of Tenure*

Both are issued by the Department of Employment and are available at council offices, Citizens' Advice Bureau and TEC offices. Some authorities run training schemes in conjunction with local industry and educational establishments.

Point of contact: usually the Planning Department – ask for the Industrial Development or Economic Development Officer.

Department of Trade and Industry

The services formally provided by the Department are now increasingly being provided by Business Link. The Department can still, however, provide useful information on available grants for start-ups.
Point of contact: telephone 0207-215 5000 and ask for the address and telephone number of the nearest DTI office and copies of their explanatory booklets.

Department of Transport and the Regions

Regulations are now in force relating to all forms of waste other than normal household rubbish. Any business that produces, stores, treats, processes, transports, recycles or disposes of such waste has a 'duty of care' to ensure it is properly discarded and dealt with.

Practical guidance on how to comply with the law (it is a criminal offence punishable by a fine not to) is contained in a booklet *Waste Management: The Duty of Care: A Code of Practice* obtainable from HMSO Publication Centre, PO Box 276, London SW8 5DT. Telephone 0207-873 9090.

Accountant

The services of an accountant are to be strongly recommended from the beginning because the legal and taxation requirements start immediately and must be properly complied with if trouble is to be avoided later. A qualified accountant must be used if a limited company is being formed but an accountant will give advice on a whole range of business issues including book-keeping, tax planning and compliance to finance raising and will help in preparing annual accounts.

It is worth spending some time finding an accountant who has other clients in the same line of business and is able to give sound advice particularly on taxation and business finance and is not so overworked that damaging delays in producing accounts are likely to arise. Ask other traders whether they can recommend their own accountant. Visit more than one firm of accountants, ask about the fees they charge and how much the production of annual accounts and agreement with the Inland Revenue are likely to cost. A good accountant is worth every penny of his fees and will save you both money and worry.

Solicitor

Many businesses operate without the services of a solicitor but there are a number of occasions when legal advice should be sought. In particular, no one should sign a lease of premises without taking legal advice because a business can encounter financial difficulty through unnoticed liabilities in its lease. Either an accountant or solicitor will help with drawing up a partnership agreement that all partnerships should have. A solicitor will also help to explain complex contractual terms and prepare draft contracts if the type of business being entered into requires them.

Insurance broker

Policies are available to cover many aspects of business including:

- employer's liability – compulsory if the business has employees
- public liability – essential in the construction industry
- motor vehicles
- theft of stock, plant and money
- fire and storm damage
- personal accident and loss of profits
- key man cover.

Brokers are independent advisers who will obtain competitive quotations on your behalf. See more than one broker before making a decision – their advice is normally given free and without obligation.
Point of contact: telephone directory or write for a list of local members to:

The British Insurance Brokers' Association
  Consumer Relations Department
  BIBA House
  14 Bevis Marks
  London
  EC3A 7NT (telephone: 0207-623 9043)

or contact
The Association of British Insurers
51 Gresham Street
London
EC2V 7HQ (telephone: 0207-600 3333)

who will supply free a package of very useful advice files specially designed for the small business.

The Health and Safety Executive

The Executive operates the legislation covering everyone engaged in work activities and has issued a very useful set of '*Construction Health Hazard Information Sheets*' covering such topics as handling cement, lead and solvents, safety in the use of ladders, scaffolding, hoists, cranes, flammable liquids, asbestos, roofs and compressed gases etc. A pack of these may be obtained free from your local HSE office or The Health & Safety Executive Central Office, Sheffield (telephone: 01142-892345) or HSE Publications (telephone: 01787-881165).

Business plan

As stated before, once the relevant information has been obtained it should be consolidated into a formal business plan. The complexity of the plan will depend in the main on the size and nature of the business concerned. Consideration should be given to the following points.

Objectives

It is important to establish what you are trying to achieve both for you and the business. A provider of finance may be particularly influenced by your ability to achieve short- and medium-term goals and may have confidence in continuing to provide finance for the business. From an individual point of view, it is important to establish goals because there is little point in having a business that only serves to achieve the expectations of others whilst not rewarding the would-be businessman.

History

If you already own an existing business then commentary on its existing background structure and history to date can be of assistance. There is no substitute for experience and any existing contacts you have in the construction industry will be of assistance to you. The following points should also be considered for inclusion:

- a brief history of the business identifying useful contacts made
- the development of the business, highlighting significant successes and their relevance to the future
- principal reasons for taking the decision to pursue this new venture
- details of present financing of the business.

Products or services

It is important to establish precisely what it is you are going to sell. Does the product or service have any unique qualities which gives it your advantages over competitors? For example, do you have an ability to react more quickly than your competitors and are you perceived to deliver a higher quality product or service? A typical business plan would include:

- description of the main products and services
- statement of disadvantages and advising how they will be overcome
- details of new products and estimated dates of introduction

- profitability of each product
- details of research and development projects
- after-sales support.

Markets and marketing strategy

This section of the business plan should show that thought has been given to the potential of the product. In this regard it can often be useful to identify major competitors and make an overall assessment of their strengths and weaknesses, including the following:

- an overall assessment of the market, setting out its size and growth potential
- a statement showing your position within the market
- an identification of main customers and how they compare
- details of typical orders and buying habits
- pricing strategy
- anticipated effect on demand of pricing
- expectation of price movement
- details of promotions and advertising campaigns.

It is important to identify your customers and why they might buy from you. Those entering the domestic side of the business will need to think about the best way to reach potential customers. Are local word-of-mouth recommendations enough to provide reasonable work continuity. If not, what is the most effective method of advertising to reach your customer base?

Remember, advertising is costly. It is a waste of funds to place an advertisement in a paper circulating in areas A, B, C & D if the business only covers area A.

Research and development

If you are developing a product or a particular service, then an assessment should be made on what stage it is at and what further finance is required to complete it. It may also be useful to make an assessment on the vulnerability of the product or service to innovations being initiated by others.

Basis of operation

Detail what facilities you will require in order to carry on your trade in the form of property, working and storage areas, office space, etc. An assessment should also he made on the assistance you will require from others. Your business plan might include:

- a layman's guide to the process or work
- details of facilities, buildings and plant
- key factors affecting production, such as yields and wastage
- raw material demand and usage.

Management

This section is one of the most important because it demonstrates the capability of the would-be businessman. The skills you need will cover production, marketing, finance and administration. In the early stages you may be able to do this yourself but as the business grows it may be required to develop a team to handle these matters. The following points should be considered for inclusion in the plan:

- set out age, experience and achievements
- state additional management requirements in the future and how they are to be met
- identify current weaknesses and how they will be overcome
- state remuneration packages and profit expectations
- give detailed CVs in appendices.

Advertising and retraining may be required in order to identify and provide suitable personnel where expertise and experience are lacking.

Financial information

It is important to detail, if any, the present financial position of your business and the budgeted profit and loss accounts, cash flows and balance sheets. These integrated forecasts should be prepared for the next twelve months at monthly intervals and annually for the following two years.

If the forecasts are to be reasonably accurate then the businessman must make some early decisions about:

- the premises where the business will be based, the initial repairs and alterations that might he required and an assessment of the total cost
- which plant, equipment and transport are needed, whether they are to be leased or purchased and what the cost will be?
- how much stock of materials, if any, should be carried? the bare minimum only should be acquired, so reliable suppliers should be found
- what will be the weekly bills for overheads, wages and the proprietor's living costs?
- what type of work is going to be undertaken, and how much profit can realistically be obtained?
- how often are invoices to be presented?

Your business plan should include the following information:

- explanation of how sales forecasts are prepared
- levels of production
- details of major variable overheads and estimates
- assumptions in cash flow forecasting, inflation and taxation.

Finance required and its application

The financial details given above should produce an accurate assessment of the funds required to finance the business. It is important to distinguish between those items that require permanent finance and those that will eventually be converted to cash because it is not usually advisable to finance long-term assets with personal equity.

Working capital such as stock and debtors can usually be obtained by an overdraft arrangement but your accountant or bank will advise you on this.

Executive summary

Although it is prepared last, this summary will be the first part of your business plan. Remember that business plans are prepared for busy people and their decision on finance may be based solely on this section. It should cover two or three pages and deal with the most important aspects and opportunities in your plan. Here are some of the main headings:

- key strategies
- finance required and how it is to be used
- management experience
- anticipated returns and profits
- markets.

The appendices should include:

- CVs of key personnel
- organisation charts
- market studies
- product advertising literature
- professional references
- financial forecasts
- glossary of terms.

If you feel that any additional information should be provided in support of your proposal, then this is usually best included in the appendices.

Follow up

Please remember that once your plan is prepared, it is important to examine it again regularly and update the forecasts and financial information. This is a working document and can be an important tool in running the business.

Sources of finance

Personal funds

Finance, like charity, often begins at home and a would-be businessman should make a realistic assessment of his net worth, including the value of his house after deducting the mortgage(s) outstanding on it, savings, any car or van owned and any sums which the family are prepared to contribute but deducting any private borrowings which will come due for payment. The whole of these funds may not be available (for instance, money which has been loaned to a friend or relative who is known to be unable to repay at the present time).

It may not be desirable that all capital should be put at risk on a business venture so the following should be established:

- how much cash you propose to invest in the business
- whether the family home will be made available for any business borrowing
- state total finance required
- how finance is anticipated being raised
- interest and security to be provided
- expected return on investment.

Whilst it may be wise not to pledge too much of the family assets, it has to be remembered that the bank will be looking closely at the degree to which the proprietor has committed himself to the venture and will not be impressed by an application for a loan where the applicant is prepared to risk only a small fraction of his own resources.

Having decided how much of his own funds to contribute, the businessman can now see the level of shortfall and consider how best to fill it. Consideration should be given to partners where the shortfall is large and particularly when there is a need for heavy investment in fixed assets, such as premises and capital equipment. It may be worthwhile starting a limited company with others also subscribing capital and to allow the banks to take security against the book debts.

Banks

The first outside source of money to which most businessmen turn is the bank and here are a few guidelines on approaching a bank manager:

- present your business plan to him; remember to use conservative estimates which tend to understate rather than overstate the forecast sales and profits
- know the figures in detail and do not leave it to your accountant to explain them for you. The bank manager is interested in the businessman not his advisers and will be impressed if the businessman demonstrates a grasp of the financing of his business

- understand the difference between short- and long-term borrowing
- ask about the Government Loan Guarantee Scheme if there is a shortage of security for loans. The bank may be able to assist, or depending on certain conditions being met, the Government may guarantee a certain percentage of the bank loan.

Remember the bank will want their money back, so bank borrowings are usually required to be secured by charges on business assets. In start-up situations, personal guarantees from the proprietors are normally required. Ensure that if these are given they are regularly reviewed to see if they are still required.

Enterprise Investment Scheme – business angels

If an outside investor is sought in a business he will probably wish to invest within the terms of the Enterprise Investment Scheme which enables him to gain income tax relief at 20% on the amount of his investment. Additionally, any investment can be used to defer capital gains tax. The rules are complex and professional advice should always be sought.

Hire purchase/leasing

It is not always necessary to purchase assets outright that are required for the business and leasing and hire purchase can often form an integral part of a business's medium-term finance strategy.

Venture capital

In addition, there are a number of other financial institutions in the venture capital market that can help well-established businesses, usually limited companies, who wish to expand. They may also assist well-conceived start-ups. They will provide a flexible package of equity and loan capital but only for large amounts, usually sums in excess of £150,000 and often £250,000.

Usually the deal involves the financial institution having a minority interest in the voting share capital and a seat on the board of the company. Arrangements for the eventual purchase of the shares held by the finance company by the private shareholders are also normally incorporated in the scheme.

The Royal Jubilee and Princes Trust

These trusts through the Youth Business Initiative provide bursaries of not more than £1,000 per individual to selected applicants who are unemployed and age 25 or over. Grants may be used for tools and equipment, transport, fees, insurance, instruction and training but not for working capital, rent and rates, new materials or stock. They operate through a local representative whose name and address may be ascertained by contacting the Prince's Youth Business.
Point of contact: telephone 0207-321 6500.

The Business Start-up Scheme

This is an allowance of £50 per week, in addition to any income made from your business, paid for twenty weeks. To qualify you must be at least 18 and under 65, work at least 36 hours per week in the business and have been unemployed for at least six months or fall into one of the other categories: disabled, ex-HMS or redundant.

The first step is to get the booklet on the subject from your local Jobcentre or TEC that includes details on how and where to apply. Once in receipt of the enterprise allowance, you will also have the benefit of advice and assistance from an experienced businessman from your TEC. All the initial counselling services and training courses are free.

RUNNING A BUSINESS

Many businesses are run without adequate information being available to check trend in their vital areas, e.g. marketing, money and managerial efficiency. It is essential to look critically at all aspects of the business in order to maximise profits and reduce inefficiency. Regular meaningful information is required on which management can concentrate. This will vary according to the proprietor's business but will often concentrate on debtors, creditors, cash, sales and orders.

Proprietors often have the feeling that the business should be 'doing better' but are unable to identify what is going wrong. Sometimes there is the worrying phenomenon of a steadily increasing work programme coupled with a persistently reducing bank balance or rising overdraft. Some useful ways of checking the position and of identifying problem areas are given below.

Marketing

Throughout his business life the entrepreneur should continuously study the methods and approach of his competitors. A shortcoming frequently found in ailing concerns is that the proprietor thinks he knows what his customers want better than they do.

The term 'market research' sounds both difficult and expensive but a very simple form of it can be done quite effectively by the businessman and his sales staff. Existing and prospective customers should be approached and asked what they want in terms of price, quality, design, payment terms, follow-up service, guarantees and services.

The initial approach might be by a leaflet or letter followed by a personal call. As an on-going part of management, all staff with customer contact should be encouraged to enquire about and record customer preferences, complaints, etc. and feed it back to management.

Other sources of information can be trade and business journals, trade exhibitions, suppliers and representatives from which information about trends, new techniques and products can be obtained and studied. Valuable information can also be gained from studying competitors and the following questions should be asked:

- what do they sell and at what prices?
- what inducements do they offer to their customers, e.g. credit facilities, guarantees, free offers and discounts?
- how do they reach their customers - local/national advertising, mail shots, salesmen, local radio and TV?
- what are the strongest aspects of their appeal to customers and have they any weaknesses?

The businessman should apply all the information gathered from customers and competitors to his own services with a view to making sure he is offering the right product at the right price in the most attractive way and in the most receptive market.

In a small business where the proprietor is also his own salesman he must give careful thought on how he can best present his product and himself. For instance, if he is working solely within the construction industry his main problems are likely to centre on getting a C1S6 Certificate and using trade contacts to get sub-contract work.

However, for those who serve the general public, presentation can be a vital element in getting work. The customer is looking for efficiency, reliability and honesty in a trader and quality, price and style in the product. To bring out these facets in discussion with a potential customer is a skilled task. A short course on marketing techniques could pay handsome dividends. The Business Link will give the names and addresses of such courses locally.

Financial control

Unfortunately, some unsuccessful firms do not seek financial advice until too late when the downward trend cannot be halted. Earlier attention to the problems may have saved some of them so it is important to recognise the tell-tale signs. There are some tests and checks that can be done quite easily.

Cash flow

Cash flow is the lifeblood of the business and more businesses fail through lack of cash than for any other reason. Cash is generated through the conversion of work into debtors and then into payment and also through the deferral of the payment of supplies for as long a period that can be negotiated. The objective must be to keep stock, work in progress, debts to a minimum and creditors to a maximum.

Debtor days

This is calculated by dividing your trade debtors by annual sales and multiplying by 365. This shows the number of days' credit being afforded to your customers and should be compared both with your normal trade terms and the previous month's figures. Normal procedures should involve the preparation of a monthly-aged list of debtors showing the name of the customer, the value and to which month it relates.

The oldest and largest debtors can be seen at a glance for immediate consideration of what further recovery action is needed. The list may also show over-reliance on one or two large customers or the need to stop supplying a particularly bad payer until his arrears have been reduced to an acceptable level. Consideration should be given to making up bills to a date before the end of the month and making sure the accounts are sent out immediately, followed by a statement four weeks later.

Consider giving discounts for prompt payment. If all else fails, and legal action for recovery is being contemplated, call at the County Court and ask for their leaflets.

Stock turn

The level of stock should be kept to a minimum and the number of days' stock can be calculated by dividing the stock by the annual purchases and multiplying by 365. A worsening trend on a month-by-month basis shows the need for action. It is important to regularly make a full inventory of all stock and dispose of old or surplus items for cash. A stock control procedure to avoid stock losses and to keep stock to a minimum should be implemented.

Profitability

Whilst cash is vital in the short-term, profitability is vital in the medium-term. The two key percentage figures are the gross profit percentage and the net profit percentage. Gross profit is calculated by deducting the cost of materials and direct labour from the sales figures whilst net profit is arrived at after deducting all overheads. Possible reasons for changes in the gross profit percentage are:

- not taking full account of increases in materials and wages in the pricing of jobs
- too generous discount terms being offered
- poor management, over-manning, waste and pilferage of materials
- too much down-time on equipment which is in need of replacement.

If net profit is deteriorating after the deduction of an appropriate reward for your own efforts, including an amount for your own personal tax liability, you should review each item of overhead expenditure in detail asking the following questions:

- can savings be made in non-productive staff?
- is sub-contracting possible and would it be cheaper?
- have all possible energy-saving methods been fully explored?
- do the company's vehicles spend too much time in the yard and can they be shared or their number reduced?
- is the expenditure on advertising producing sales - review in association with 'marketing' above?

Over-trading

Many inexperienced businessmen imagine that profitability equals money in the bank and in some cases, particularly where the receipts are wholly in cash, this may be the case. But often, increased business means higher stock inventories, extra wages and overheads, increased capital expenditure on premises and plant, all of which require short-term finance.

Additionally, if the debtors show a marked increase as the turnover rises, the proprietor may find to his surprise that each expansion of trade reduces rather than increases his cash resources and he is continually having to rely on extensions to his existing credit.

The business, which had enough funds for start-up, finds it does not have sufficient cash to run at the higher level of operation and the bank manager may he getting anxious about the increasing overdraft. It is essential for those who run a business that operates on credit terms to be aware that profitability does not necessarily mean increased cash availability. Regular monthly management information on marketing and finance as described in this chapter will enable over-trading to be recognised and remedial action to be taken early.

If the situation is appreciated only when the bank and other creditors are pressing for money, radical solutions may be necessary, such as bringing in new finance, sale and leaseback of premises, a fundamental change in the terms of trade or even selling out to a buyer with more resources. Help from the firm's accountant will be needed in these circumstances.

Break-even point

The costs of a business may be divided into two types – variable and fixed. *Variable costs* are those which increase or decrease as the volume of work goes up or down and include such items as materials used, direct labour and power machine tools. *Fixed costs* are not related to turnover and are sometimes called fixed overheads. They include rent, rates, insurance, heat and light, office salaries and plant depreciation. These costs are still incurred even though few or no sales are being made.

Many small businessmen run their enterprises from home using family labour as back-up; they mainly sell their own labour and buy materials and hire plant only as required. By these means they reduce their fixed costs to a minimum and start making profits almost immediately. However, larger firms that have business premises, perhaps a small workshop, an office and vehicles, need to know how much they have to sell to cover their costs and become profitable.

In the case of a new business it is necessary to estimate this figure but where annual accounts are available a break-even chart based on them can be readily prepared. Suppose the real or estimated figures (expressed in £000s) are:

	%	£
Sales	100	400
Variable costs	66	265
Gross profit	34	135
Fixed costs	13	50
Net profit	21	85

Break-even point   =   <u>50 divided by (1 less variable costs %)</u>
                                                   sales

                                    =   50 divided by (1 less 0.6625)
                                    =   50 divided by 0.3375
                                    =   £148 (thousand)

In practice, things are never quite as clear cut as the figures show, but nevertheless this is a very useful tool for assessing not only the break-even point but also the approximate amount of loss or profit arising at differing levels of turnover and also for considering pricing policy.

TAXATION

The first decision usually required to be made from a taxation point of view is which trading entity to adopt. The options available are set out below.

Sole trader

A sole trader is a person who is in business on his own account. There is no statutory requirement to produce accounts nor is there a necessity to have them audited. A sole trader may, however, be required to register for PAYE and VAT purposes and maintain records so that Income Tax and VAT returns can be made. A sole trader is personally liable for all the liabilities of his business.

Partnership

A partnership is a collection of individuals in business on their own account and whose constitution is generally governed by the Partnership Act 1890. It is strongly recommended that a partnership agreement is also established to determine the commercial relationship between the individuals concerned.

The requirements in relation to accounting records and returns are similar to those of a sole trader and in general a partner's liability is unlimited.

Limited company

This is the most common business entity. Companies are incorporated under the Companies Act 1985 which requires that an annual audit is carried out for all companies with a turnover in excess of £5,000,000 or a review if the turnover is less than £5,000,000 and that accounts are filed with the Companies Registrar. Generally an individual shareholder's liability is limited to the amount of the share capital he is required to subscribe.

Advantages

In view of the problems and costs of incorporating an existing business, it is important to try and select the correct trading medium at the commencement of operations. It is not true to say that every business should start life as a company.

Many businesses are carried on in a safe and efficient manner by sole traders or partnerships. Whilst recognising the possible commercial advantages of a limited company, taxation advantages exist for sole traderships and partnerships, such as income tax deferral and National Insurance saving. No decision should be taken without first seeking professional advice.

The benefit of limited liability should not be ignored although this can largely be negated by banks seeking personal guarantees. In addition, it may be easier for the companies to raise finance because the bank can take security on the debts of the company that could be sold in the future, particularly if third-party finance has been obtained in the form of equity.

Self-assessment

From the tax year 1996/97 the burden of assessing tax shifted from the Inland Revenue to the individual tax payer. The main features of this system are as follows:

-	the onus is on the taxpayer to provide information and to complete returns
-	tax will be payable on different dates
-	the taxpayer has a choice: he can calculate his tax liability at the same time as making his return and this will need to be done by 31st January following the end of the tax year. Alternatively, he can send in his tax return before 30 September and the Inland Revenue will calculate the tax to be paid on the following 31 January
-	the important aspect to the system is that if the return is late, or the tax is paid late, there will be automatic penalties and/or surcharges imposed on the taxpayer.

Tax correspondence

Businessmen do not like letters from the Inland Revenue but they should resist the temptation to tear them up or put them behind the clock and forget about them. All Tax Calculations and Statements of Account should be checked for accuracy immediately and any queries should be put to your accountant or sent to the Tax District that issued the document.

Keep copies of all correspondence with the Inland Revenue. Letters can be mislaid or fail to be delivered and it is essential to have both proof of what was sent as well as a permanent record of all correspondence.

Dates tax due

Income Tax
Payments on account (based on one half of last year's liability) are due on 31 January and 31 July. If these are insufficient there is a balancing payment due on the following 31 January – the same day as the tax return needs to be filed. For example:

> for the year 2006/07   Tax due £5,000 (2005/06 was £4,000)
> First payment on account of £2,000 is due on 31.01.07
> Second payment on account of £2,000 is due on 31.07.07
> Balancing payment of £1,000 is due on 31.01.08

Note that on 31.01.07, the first payment on account of £2,500 fell due for the tax year 2007/08.

Tax in business

Spouses in business

If spouses work in the business, perhaps answering the phone, making appointments, writing business letters, making up bills and keeping the books, they should be properly remunerated for it. Being a payment to a family member, the Inspector of Taxes will be understandably cautious in allowing remuneration in full as a business expense. The payment should be:

- actually paid to them, preferably weekly or monthly and in addition to any housekeeping monies
- recorded in the business book

- reasonable in amount in line with their duties and the time spent on them.

If the wages paid to them exceed £96.00 per week, Class 1 employer's and employee's NIC becomes due and if they exceed £5,225 p.a. (assuming they have no other income) PAYE tax will also be payable.

It should also be noted that once small businesses are well established and the spouses' earnings are approaching the above limits, consideration may be given to bringing them in as a partner. This has a number of effects:

- there is a reduced need to relate the spouse's income (which is now a share of the profits) to the work they do
- they will pay Class 2 and Class 4 NIC instead of the more costly Class I contributions and PAYE will no longer apply to their earnings but remember that, as partners, they have unlimited liability.

Premises

Many small businessmen cannot afford to rent or buy commercial premises and run their enterprises from home using part of it as an office where the books and vouchers, clients' records and trade manuals are kept and where estimates and plans are drawn up. In these circumstances, a portion of the outgoings on the property may be claimed as a business expenses. An accountant's advice should be sought to ensure that the capital gains tax exemption that applies on the sale of the main residence is not lost.

Fixed Profit Car Scheme

It may be advantageous to calculate your car expenses using a fixed rate per business mile. Ask your accountant about this. A proper record of business mileage must be kept.

Vehicles

Car expenses for sole traders and partners are usually split on a fractional mileage basis between business journeys, which are allowable, and private ones, which are not, and a record of each should he kept. If the business does work only on one or two sites for only one main contractor, the inspector may argue that the true base of operations is the work site not the residence and seek to disallow the cost of travel between home and work. It is tax-wise and sound business practice to have as many customers as possible and not work for just one client.

Business entertainment

No tax relief is due for expenditure on business entertainment and neither is the VAT recoverable on gifts to customers, whether they are from this country or overseas. However, the cost of small trade gifts not exceeding £50 per person per annum in value is still admissible provided that the gift advertises the business and does not consist of food, drink or tobacco.

Income tax (2006/07)

Personal allowances

The current personal allowance for a single person is £5,035. The personal allowance for people aged 65 to 74 and over 75 years are £7,280 and £7,420 respectively. The married couple's allowance was withdrawn on 5 April 2000, except for those over 65 on that date.

Taxation of husband and wife

A married woman is treated in much the same way as a single person with her own personal allowance and basic rate band. Husband and wife each make a separate return of their own income and the Inland Revenue deals with each one in complete privacy; letters about the husband's affairs will be addressed only to him and about the wife's only to her unless the parties indicate differently.

Rates of tax

Tax is deducted at source from most banks and building societies accounts at the rate of 20%. The rates of tax for 2006/07 are as follows:

Lower rate: 10% on taxable income up to £2,150
Basic rate: 22% on taxable income between £2,150 and £33,300
Higher rate: 40% on taxable income over £33,300

Dividends carry a 10% non-repayable tax credit. Higher rate taxpayers pay a further tax on dividends of 22.5%.

Mortgage interest relief

This is no longer available after 5 April 2000.

Business losses

These are allowed only against the income of the person who incurs the loss. For example, a loss in the husband's business cannot be set against the wife's income from employment.

Joint income

In the case of joint ownership by a husband and wife of assets that yield income, such as bank and building society accounts, shares and rented property, the Inland Revenue will treat the income as arising equally to both and each will pay tax on one half of the income. If, however, the asset is owned in unequal shares or one spouse only and the taxpayer can prove this, then the shares of income to be taxed can be adjusted accordingly if a joint declaration is made to the tax office setting out the facts.

Capital Gains Tax

Where an asset is disposed of, the first £8,800 of the gain is exempt from tax. In the case of husbands and wives, each has a £8,800 exemption so if the ownership of the assets is divided between them, it is possible to claim exemption on gains up to £17,600 jointly in the tax year. Any remaining gain is chargeable as though it were the top slice of the individual's income; therefore according to his or her circumstances it might be charged at 10%, 22% or 40%.

Self-employed NIC rates (from 6 April 2006)

Class 2 rate
Charged at £2.10 per week. If earnings are below £5,035 per annum averaged over the year, ask the DSS about 'small income exception'. Details are in leaflet CA02.

Class 4 rate
Business profits up to £5,035 per annum are charged at NIL. Annual profits between £5,035 and £33,540 are charged at 8% of the profit. There is also a charge on profits over £33,540 of 1%. Class 4 contributions are collected by the Inland Revenue along with the income tax due.

Capital allowances (depreciation) rates

Plant and machinery:	25% (40% first-year allowance is available for certain small businesses)
Business motor cars − cost up to £12,000:	25%
− cost over £12,000:	£3,000 (maximum)

THE CONSTRUCTION INDUSTRY TAX DEDUCTION SCHEME

General

The new Construction Industry Tax Deduction Scheme is known as the 'revised CIS' scheme and replaced the old 'CIS' scheme. Everyone who carries out work in the Construction Industry Scheme must hold a registration card (CIS4) or a tax certificate (CIS6). Certain larger companies use a special certificate (CIS5).

A small business that does work only for the general public and small commercial concerns is outside the scheme and does not need a certificate to trade. If, however, it engages other contractors to do jobs for it, the business would have to trade under the scheme as a contractor and deduct tax from any payment made to a subcontractor who did not produce a valid (CIS6) certificate. If in doubt, consult your accountant or the Inland Revenue direct.

Under the revised scheme, registration cards, tax certificates and vouchers will no longer be used. Subcontractors will normally register with the Revenue. There will be two types of registration, registration for gross payment, applicable to those who previously qualified for a tax exemption certificate, and registration for payment under deduction, applicable to all other subcontractors. The three-year qualifying period for gross payment will be reduced to one year (but the Revenue may cancel registration at any time where the qualifying conditions no longer apply, or where the rules have been breached). Unlike the present scheme contractors will be able to pay un-registered sub-contractors, but the deduction rate will be much higher (probably around 30%).Those who hold a subcontractor's certificate when the new provisions come into effect will be treated as being registered for gross payment and those who hold a registration card will be treated as registered for payment for deduction. Subcontractors holding temporary registration that expire before 6 April 2007 will have to register in the same way as new applicants.

For new workers, contractors will need to obtain basic identity details and will check with the Revenue what type of registration the worker holds. The verification may be done by telephone or over the internet. Contractors may assume that the status remains unchanged unless the Revenue notifies them to the contrary.

When contractors deduct tax from payments, they must supply the subcontractors with pay statements. Contractors will no longer be required to submit payment vouchers monthly. Instead, they will be required to submit monthly returns (even for those who make payments quarterly) and must provide details of recipients and payments made, together with a 'status declaration' that none of the payments relate to a contract of employment . This requirement puts the onus of establishing the worker's status on the contractor and there will be penalties for false declarations (and also various other penalties, as now, for both contractors and workers). The returns must be sent in every month, even if no payments have been made, and penalties will apply if a return is not submitted. Nil returns will be able to be made on paper, over the internet or by telephone. It is intended that other online services, such as subcontractor verification checks, will also be available.

VAT

The general rule about liability to register for VAT is given in the VAT office notes. It is possible to give here only a brief outline of how the tax works. The rules that apply to the construction industry are extremely complex and all traders must study *The VAT Guide* and other publications.

Registration for VAT is required if, at the end of any month, the value of taxable supplies in the last 12 months exceeds the annual threshold or if there are reasonable grounds for believing the value of the taxable supplies in the next 30 days will exceed the annual threshold.

Taxable supplies include any zero-rated items. The annual threshold is £64,000. The amount of tax to be paid is the difference between the VAT charged out to customers (*output tax*) and that suffered on payments made to suppliers for goods and services (*input tax*) incurred in making taxable supplies. Unlike income tax there is no distinction in VAT for capital items so that the tax charged on the purchase of, for example, machinery, trucks and office furniture, will normally be reclaimable as *input tax*.

VAT is payable in respect of three monthly periods known as 'tax periods'. You can apply to have the group of tax periods that fits in best with your financial year. The tax must be paid within one month of the end of each tax period. Traders who receive regular repayments of VAT can apply to have them monthly rather than quarterly. Not all types of goods and services are taxed at 17.5% (i.e. the standard rate). Some are exempt and others are zero-rated.

Zero-rated

This means that no VAT is chargeable on the goods or services, but a registered trader can reclaim any *input* tax suffered on his purchases. For instance, a builder pays VAT on the materials he buys to provide supplies of constructing but if he is constructing a new dwelling house, this is zero rated. The builder may reclaim this VAT or set it off against any VAT due on standard rated work.

Exempt

Supplies that are exempt are less favourably treated than those that are zero rated. Again no VAT is chargeable on the goods or services but the trader cannot reclaim any *input* tax suffered on his purchases.

Standard-rated

All work which is not specifically stated to be zero rated or exempt is standard-rated, i.e. VAT is chargeable at the current rate of 17.5% and the trader may deduct any *input* tax suffered when he is making his return to the Customs and Excise. If for any reason a trader makes a supply and fails to charge VAT when he should have done so (e.g. mistakenly assuming the supply to be zero rated), he will have to account for the VAT himself out of the proceeds. If there is any doubt about the VAT position, it is safer to assume the supply is standard rated, charge the appropriate amount of VAT on the invoice and argue about it later.

Time of supply

The *time* at which a supply of goods or services is treated as taking place is important and is called the 'tax point'. VAT must be accounted for to the Customs and Excise at the end of the accounting period in which this 'tax point' occurs. For the supply of goods which are 'built on site', the 'basic tax point' is the date the goods are made available for the customer's use, whilst for *services* it is normally the date when all work except invoicing is completed.

However, if you issue a tax invoice or receive a payment before this 'basic tax point' then that date becomes a tax point. In the case of contracts providing for stage and retention payments, the tax point is either the date the tax invoice is issued or when payment is received, whichever is the earlier.

All the requirements apply to sub-contractors and main contractors and it should be noted that, when a contractor deducts income tax from a payment to a sub-contractor (because he has no valid CIS6) VAT is payable on the full gross amount *before* taking off the income tax.

Annual accounting

It is possible to account for VAT other than on a specified three month period. Annual accounting provides for nine equal installments to be paid by direct debit with annual return provided with the tenth payment. £300,000.

Cash accounting

If turnover is below a specified limit, currently £660,000, a taxpayer may account for VAT on the basis of cash paid and received. The main advantages are automatic bad debt relief and a deferral of VAT payment where extended credit is given.

Bad debts

Relief is available for debts over 6 months.

Part Seven

GENERAL CONSTRUCTION DATA

General construction data

GENERAL CONSTRUCTION DATA

The metric system

Linear

1 centimetre (cm)	=	10 millimetres (mm)
1 decimetre (dm)	=	10 centimetres (cm)
1 metre (m)	=	10 decimetres (dm)
1 kilometre (km)	=	1000 metres (m)

Area

100 sq millimetres	=	1 sq centimetre
100 sq centimetres	=	1 sq decimetre
100 sq decimetres	=	1 sq metre
1000 sq metres	=	1 hectare

Capacity

1 millilitre (ml)	=	1 cubic centimetre (cm3)
1 centilitre (cl)	=	10 millilitres (ml)
1 decilitre (dl)	=	10 centilitres (cl)
1 litre (l)	=	10 decilitres (dl)

Weight

1 centigram (cg)	=	10 milligrams (mg)
1 decigram (dg)	=	10 centigrams (mcg)
1 gram (g)	=	10 decigrams (dg)
1 decagram (dag)	=	10 grams (g)
1 hectogram (hg)	=	10 decagrams (dag)

Conversion equivalents (imperial/metric)

Length

1 inch	=	25.4 mm
1 foot	=	304.8 mm
1 yard	=	914.4 mm
1 yard	=	0.9144 m
1 mile	=	1609.34 m

Area

1 sq inch	=	645.16 sq mm
1 sq ft	=	0.092903 sq m
1 sq yard	=	0.8361 sq m
1 acre	=	4840 sq yards
1 acre	=	2.471 hectares

Liquid

1 lb water	=	0.454 litres
1 pint	=	0.568 litres
1 gallon	=	4.546 litres

Horse-power

1 hp	=	746 watts
1 hp	=	0.746 kW
1 hp	=	33,000 ft.lb/min

Weight

1 lb	=	0.4536 kg
1 cwt	=	50.8 kg
1 ton	=	1016.1 kg

Conversion equivalents (metric/imperial)

Length

1 mm	=	0.03937 inches
1 centimetre	=	0.3937 inches
1 metre	=	1.094 yards
1 metre	=	3.282 ft
1 kilometre	=	0.621373 miles

Area

1 sq millimetre	=	0.00155 sq in
1 sq metre	=	10.764 sq ft
1 sq metre	=	1.196 sq yards
1 acre	=	4046.86 sq m
1 hectare	=	0.404686 acres

Weight

1 kg	=	2.205 lbs
1 kg	=	0.01968 cwt
1 kg	=	0.000984 ton

Temperature equivalents

In order to convert Fahrenheit to Celsius deduct 32 and multiply by 5/9.
To convert Celsius to Fahrenheit multiply by 9/5 and add 32.

Fahrenheit	Celsius
230	110.0
220	104.4
210	98.9
200	93.3
190	87.8
180	82.2
170	76.7
160	71.1
150	65.6
140	60.0
130	54.4
120	48.9
110	43.3
100	37.8
90	32.2
80	26.7
70	21.1
60	15.6
50	10.0
40	4.4
30	-1.1
20	-6.7
10	-12.2
0	-17.8

Areas and volumes

Figure	Area	Perimeter
Rectangle	Length × breadth	Sum of sides
Triangle	Base × half of perpendicular height	Sum of sides
Quadrilateral	Sum of areas of contained triangles	Sum of sides
Trapezoidal	Sum of areas of contained triangles	Sum of sides
Trapezium	Half of sum of parallel sides × perpendicular height	Sum of sides
Parallelogram	Base × perpendicular height	Sum of sides
Regular polygon	Half sum of sides × half internal diameter	Sum of sides
Circle	pi × radius^2	pi × diameter or pi × 2 × radius

Figure	Surface area	Volume
Cylinder	pi × 2 × radius^2 × length (curved surface only)	pi × radius^2 × length
Sphere	pi × diameter^2	Diameter^3 × 0.5236

Weights of materials	kg/m^2
Plaster	
Carlite browning. 11mm thick	7.80
Carlite tough coat, 11mm thick	7.80
Carlite bonding, 8mm thick	7.10
Carlite bonding, 11mm thick	9.80

Weights of materials   kg/m²

Thistle hardwall, 11mm thick	8.80
Thistle dri-coat, 11mm thick	8.30
Thistle renovating, 11mm thick	8.80

PLASTERING AND TILING

Plaster coverage   m² per
                                              1000kg

Carlite browning, 11mm thick	135–155
Carlite tough coat, 11mm thick	135–150
Carlite bonding, 11mm thick	100–115
Thistle hardwall, 11mm thick	115–130
Thistle dri-coat, 11mm thick	135–140
Thistle renovating, 11mm thick	115–125

Tile coverage   Nr

152 × 152mm	43.27
200 × 200mm	25.00

PAINTING AND WALLPAPERING

Average coverage of paints m² per litre	Timber	Plastered surfaces	Brickwork
Primer	10–12	9–11	5–7
Undercoat	10–12	11–14	6–8
Gloss	11–14	11–14	6–8
Emulsion	10–12	12–15	6–10

Wallpaper coverage per roll	Rolls nr	Wall height m	Room perimeter m
	4	2.50	8
	5	2.50	9
	5	2.50	10
	6	2.50	11
	6	2.50	12
	7	2.50	13
	7	2.50	14
	8	2.50	15
	8	2.50	16
	8	2.50	17
	9	2.50	18
	10	2.50	19
	10	2.50	20
	10	2.50	21
	11	2.50	22
	11	2.50	23
	12	2.50	24
	13	2.50	25
	13	2.50	26
	14	2.50	27
	5	2.80	8
	5	2.80	9
	5	2.80	10
	7	2.80	11
	7	2.80	12
	7	2.80	13
	8	2.80	14
	8	2.80	15
	9	2.80	16
	10	2.80	17
	10	2.80	18
	11	2.80	19
	11	2.80	20
	12	2.80	21

Wallpaper coverage per roll

Rolls nr	Wall height m	Room perimeter m
13	2.8	22
13	2.8	23
14	2.8	24
14	2.8	25
15	2.8	26
15	2.8	27

Index

Printed in the United States
by Baker & Taylor Publisher Services